Static
Headspace–Gas
Chromatography

Static Headspace–Gas Chromatography

Theory and Practice

Bruno Kolb

and

Leslie S. Ettre

 **WILEY-VCH**

New York · Chichester · Weinheim · Brisbane · Singapore · Toronto

CHEM

Bruno Kolb
Bodenseewerk Perkin-Elmer GmbH
Überlingen, German Federal Republic

Leslie S. Ettre
Department of Chemical Engineering
Yale University
New Haven, CT, USA

This book is printed on acid-free paper. ⊗

Library of Congress Cataloging-in-Publication Data

Kolb, Bruno.
    Static headspace-gas chromatography : theory and practice / Bruno
Kolb and Leslie S. Ettre.
        p.   cm.
    Includes bibliographical references and index.
    ISBN 1-56081-696-1 (alk. paper)
    1. Gas chromatography.   I. Ettre, Leslie S.   II. Title.
QD79.C45K64   1997   96-27616
543′.0896--dc20   CIP

Printed in the United States of America

ISBN 0-471-19238-4 Wiley-VCH, Inc.

10 9 8 7 6 5 4 3 2 1

Contents

Chapter 3   **The Technique of Headspace–Gas
Chromatography   45**

Chapter 6   Method Development in HS–GC   211

Preface

Headspace–gas chromatography is not new; the technique has been practiced since the early days of gas chromatography. However, there is still an increasing interest, apparently driven by the need for cost reduction in every analytical laboratory. This calls for automation of each part of an analytical procedure. Computer-controlled automated analytical instruments including autosamplers and data systems were the first step in this process of automation and were carried out very effectively by the instrument manufacturers. While the actual time needed to perform the analysis could thus be dramatically reduced, sample preparation remains a time-consuming task. Despite variation from laboratory to laboratory, our experience, which has been confirmed by statistical surveys, is that in most laboratories about two-thirds of the time is spent on sample preparation, while only 10% goes for actual analysis, and the rest for documentation and organization. Whenever the efficiency of an analytical laboratory has to be improved, it is worthwhile to look at and begin with sample handling.

Most samples need to be modified for the specific requirement of a particular analytical technique. Most of these cleanup procedures use some type of initial extraction procedure such as solvent extraction, solid-phase extraction, or supercritical-fluid extraction. However, if we are interested in highly volatile compounds, we can use an inert gas for this purpose; gas is an ideal "solvent" for volatile compounds, since it is easy to handle and is available in a much higher purity than most organic solvents—an aspect that is particularly important for trace analysis. A gas extract is ideally suited for analysis by gas chromatography, and this combination is called "headspace–gas chromatography"—HS-GC. Gas extraction techniques can be carried out in several variants: as a single step (static headspace) or by stepwise repeating of the extraction (multiple headspace extraction) and also by stripping the volatiles (dynamic headspace) by a continuous flow of an inert purge gas. All these gas extraction techniques are called headspace techniques for historical reasons (the name "headspace" was originally given to the gas content of the bulge that forms at the top of a can of food, whose composition had to be analyzed).

If gas extraction is considered to be a suitable cleanup procedure for a particular sample, we may ask which of these variants should finally be applied. The following criteria may help to make this decision:

- simplicity of operation
- degree of automation
- flexibility for changing requirements
- sensitivity
- quantitation

The simplicity of static HS-GC is unsurpassed by any other cleanup techniques: the sample (either a gas, a liquid, or a solid) is filled into the headspace vial, which is closed immediately and remains closed until an aliquot has been withdrawn from the closed vial and transferred directly to the gas chromatographic system, thus guaranteeing sample integrity. This simplicity enabled the early automation of the whole procedure. It is interesting to note that Bodenseewerk Perkin-Elmer introduced the first automated headspace sampler for gas chromatography as early as 1967 and prior to any liquid autosamplers. Automation also helps to overcome the only drawback of static HS-GC, the sometimes long equilibration times.

The flexibility of a system in adapting to varying sample properties is also an important factor in saving time if a laboratory receives samples of different types for analysis. This argument favors static against dynamic HS-GC: it has fewer parameters to be tailored and optimized for specific sample properties, such as selecting the various adsorbents for filling a trap in the case of dynamic HS-GC. In principle, static HS-GC needs to determine only the purely physical parameters (i.e., time and temperature) to achieve the necessary state of equilibrium in the vial.

As far as sensitivity and the possibilities for quantitative analysis are concerned, one would at first favor dynamic HS-GC. Its inherent purpose is to perform an exhaustive extraction, contrary to static HS-GC, and therefore the composition of the resulting gas extract is often considered to be the same as that of the original sample. As this book shows, however, the modern techniques of cryogenic focusing also allow the sensitivity range to be extended to determine concentrations down to the level of parts per trillion, or even parts per quadrillion (ppt, $1:10^{-12}$; ppq, $1:10^{-15}$).

With static HS-GC, the quantitative aspects are often rendered difficult, or at least complicated, by the somewhat mysterious matrix influence. A few words of clarification are necessary here. The first application of automated static HS-GC was the quantitative determination of ethanol in blood samples. No other analytical technique has been investigated and tested worldwide for precision, accuracy, reliability, and robustness by so many independent experts. If static HS-GC works so well with such a complex matrix as blood, there is no reason for it to fail in other cases and with other matrices. Therefore, it is the main concern of the authors of this book to focus on the quantitative aspects of static HS-GC.

The need to give a comprehensive discussion of all the possible calibration techniques for gas, liquid, or solid samples in HS-GC was recognized by the authors during many headspace training courses — for example, the course entitled Headspace Gas Chromatography: Equilibrium and Purge-and-Trap Analysis, which we have held for several years at the Pittsburgh Conference on Analytical Chemistry and Applied Spectroscopy. The many questions and discussions with the participants stimulated the writing of this book, which we hope will prove to be a real practical textbook. For this reason, we have included many practical examples with all the original data (peak area values), to enable the interested reader to go through all the calculations and also to use these data for alternative methods. We have compared the quantitative results of various calibration techniques as appropriate, to demonstrate that there feasible alternative techniques often exist.

Although this book emphasizes techniques, methods, and procedures rather than applications, we have selected the many practical examples to cover at least the most important applications of static HS-GC in environmental, polymer, and food analysis, and in some other interesting fields of application. Most of these applications, if not otherwise referred to literature, were developed in the GC laboratory at Bodenseewerk Perkin-Elmer, which pioneered in the automated HS-GC, most of these are unpublished results. It is natural therefore that these practical examples were carried out by the "balanced pressure headspace sampling technique," specific to the Perkin-Elmer headspace samplers. However, this is not mentioned explicitly in the figures, because a headspace sample is in fact a gas sample, and any other technique for the introduction of gas sample into a gas chromatograph should in principle also be suitable. The use of a particular sampling technique, therefore, should not be considered a biased preference.

This book would not have been possible without the great commitment of many co-workers in this GC laboratory. Of these we thank particularly Maria Auer and Petr Pospisil, who made many valuable contributions to both instrumental engineering and applications. If this book contains many useful practical hints, it is due to the highly skilled experimental work and experience of Mrs. Auer, who carried out most of the quantitative examples in this book. We also thank Meredith Harral-Schulz for preparing the manuscript and Albert Grundler for the design of many of the figures.

Bruno Kolb

Leslie S. Ettre

List of Acronyms and Symbols

In general we follow the recommendations of the Nomenclature for Chromatography* of the International Union of Pure and Applied Chemistry (IUPAC) and use their recommendations whenever possible, including the rules for how to use subscripts and superscripts.

Here we list the general symbols and acronyms used in this book. Further differentiation is carried out by the use of *subscripts*. Generally used subscripts are identified below. Additional general subscripts used include *1, 2,* and so forth for subsequent measurements; *i* to express in general a given analyte; *st* to indicate the standard; and *ex* the external standard. Subscripts *o*, *S*, and *G* refer to the original situation, the sample, and the gas phase, respectively; subscript *o* may also indicate conditions at the column outlet or a base situation. Some additional specific subscripts are explained in the respective chapters.

Based on the recommendation of IUPAC, the use of *superscripts* is avoided except when specifically distinguishing between sample and calibration measurement, where the latter is indicated by superscript *c*, and in the symbols for saturated vapor pressure (p^o) and for energy of mixing (ΔG^M).

In expressing concentration, we always use *parts-per-million* (*ppm*) and *parts-per-billion* (*ppb*). According to the American usage, *ppm* means $1:10^6$, e.g., mg/Liter, while *ppb* means $1:10^9$, e.g., μg/Liter.

Pure Appl. Chem. **65**, 819–872 (1993).

Acronyms*

AA	acetaldehyde
ASTM	*American Society for Testing & Materials*
BF	backflush technique
BP	boiling point
BTEX	benzene, toluene, ethylbenzene, and the xylenes
BVB	1-bromo-4-fluorobenzene
CEN	*Comité Européen de Normalization* (European Commitee for Standardization)
DIN	*Deutsche Industrienormen* (German Industrial Standards)
DMA	dimethyl acetamide
DMF	dimethylformamide
ECD	electron-capture detector
EG	ethylene glycol
EHA	2-ethylhexyl acrylate
ELCD	electrolytic conductivity detector
EO	ethylene oxide
EPA	*Environmental Protection Agency (U.S.A.)*
FDA	*Food and Drug Administration (U.S.A.)*
FET	full-evaporation technique
FID	flame-ionization detector
FPD	flame-photometric detector
FTIR	Fourier-transform infrared spectroscopy
GC	gas chromatography
GPA	gas-phase addition
HS	headspace
HSA	headspace analysis
HS-GC	headspace–gas chromatography
IF	improvement factor
IUPAC	*International Union of Pure and Applied Chemistry*

*Names of associations and official groups are given in italics.

KF	Karl-Fischer titration
MEK	methyl ethyl ketone
MHE	multiple headspace extraction
MHI	multiple headspace injection technique
MS	mass spectrometer
NPD	nitrogen-phosphorous detector or thermionic detector
OVIs	organic volatile impurities
PET	poly(ethylene terephthalate)
PGC	propylene glycol carbonate
PID	photo-ionization detector
PRV	phase-ratio variation method
PS	polystyrene
PVC	poly(vinyl chloride)
RF	response factor
RSD	relative standard deviation
SM	styrene monomer
SPA	sample phase addition
TCD	thermal-conductivity detector
TCE	tetrachloroethylene
TCTA	2,4,6-trichloro-1,3,5-triazine
TVT	total vaporization technique
UNIFAC	universal functional group activity coefficients
USP	U.S. Pharmacopeia
VCM	vinyl chloride monomer
VDI	*Verein Deutscher Ingenieure* (Association of German Engineers)
VOCs	volatile organic compounds
VPC	vapor-phase calibration method
WCC	whole-column cryotrapping procedure

Symbols

a, a'	constant (in general)
a, a'	constant (slope) of a linear regression equation
a_C, a_G	constant (slope) in the linear regression evaluation of the VPC method
A	peak area
A_C	peak area corresponding to C_C
A'_C	peak area corresponding to W_C in the determination of K by the VPC method
A^x	peak area corrected for sample volume in the vial
A_1	in MHE, the measured area of the first extraction
A_1^*	in MHE, the theoretical area corresponding to the first extration (the intercept values)

ΔA	peak area corresponding to an added amount of the analyte (standard addition method)
b, b'	constant (intercept) of a linear regression equation
b_C, b_G	constant (intercept) in the linear regression evaluation of the VPC method
B	constant (e.g., in Antoine-type equations)
B'	constant (e.g., in Antoine-type equations)
C	constant (e.g., in Antoine-type equations)
C'	constant (e.g., in Antoine-type equations)
c_n	number of carbons atoms in a molecule (carbon number)
C_C	concentration of the analyte in the calibration vial in the VPC method
C_o	original concentration of the analyte in the sample
C_e	actual concentration of a compound during exponential gas dilution
C_G	analyte concentration in the gas phase (headspace)
C_S	analyte concentration in the sample phase
d	density
d	diffusion path length
d_c	inside column diameter
d_f	coated stationary phase film thickness
D	diffusion coefficient
DL	minimum detectable limit
f	proportionality factor (in general)
f	calibration factor, correction factor, or response factor
f	friction factor
f_c	calibration factor
f_V	volume correction factor
F	flow rate (in general)
$F_{c,o}$	carrier gas flow rate at column outlet, corrected to vial temperature and dry gas—conditions
F_i	carrier gas flow rate at column inlet
F_a	carrier gas flow rate at column outlet
ΔG^M	total free energy of mixing
ΔG_i^M	partial free molar energy of mixing
H	peak height*
i	reference to a certain compound or to the stage in a measurement (e.g., in MHE)
K	partition (distribution) coefficient
L	column length
M	molecular weight
n	number of moles of a compound
n	number of measurements

---

*In the general chromatography nomenclature, H and N are the symbols of plate height (HETP) and plate number, respectively. In this book, however, column efficiency is never specified, so H and N are given the meanings listed here.

n_{total} total number of moles present
N noise level*
p pressure (in general)
p_a ambient pressure
p_h in MHE, pressure in the headspace vial
p_i partial pressure of a compound
p_i inlet pressure to a column (absolute)
p_L pressure in the sample loop
p^o saturation vapor pressure of a compound
p_o in MHE, pressure in the headspace vial after venting
p_o pressure at standard conditions
p_p pressurization pressure
p_{total} total pressure of a gas mixture
p_v sample vapor pressure in the headspace vial
p_V pressure in the headspace vial after pressurization
p_w partial pressure of water at ambient temperature
Δp pressure drop along the column
$p\%$ precision of a detector's linear range
q constant in the exponent describing MHE
Q area ratio of two consecutive peaks in MHE
Q_c cross section of a column
r correlation coefficient (in linear regression)
r ratio of amounts in the determination of K by the VPC method
R gas constant
R peak area ratios (in standard addition method)
RF response factor
RR rate of release
S selectivity
S_{sample} surface area of sample
t time
T absolute temperature
T_a ambient temperature
T_c column temperature
T_g glass transition temperature
T_v vial temperature
u average linear carrier gas velocity
V volume (in general)
V_e the expanded volume of the headspace gas at p_o
V_G volume of the gas phase (headspace) in the vial
V_H volume of transferred headspace gas
V_L gas volume filling a sample loop
V_{mole} g-mole volume of a pure compound in gaseous (vapor) form
V_o volume of the original sample
V_S volume of sample phase in a vial
V_V total volume of a vial

V_{vent} volume of gas vented in the MHE procedure
W amount (in general)
W_a added amount of the analyte
W_c amount added to the vial in the determination of K by the VPC method
W_A amount of the analyte in the aliquot withdrawn from the headspace
W_G amount of the analyte in the gas phase (headspace)
W_o original amount of the analyte present in the sample
W_S amount of analyte in the sample phase
x mole fraction (in general)
$x_{G,i}$ mole fraction of a component in a gas mixture
$x_{S,i}$ mole fraction of a component in solution
$Y\%_o$ extraction yield in the full-evaporation technique
α proportionality constant
β phase ratio
ϕ_s phase fraction (sample volume as a function of the vial's volume)
γ activity coefficient
η carrier gas viscosity
ρ relative pressure (p_o/p_h) in MHE

1

General Introduction

1.1 Principles of Headspace Analysis

Gas chromatography is an analytical technique for the investigation of volatile compounds. If the sample is a gas, then an aliquot of it is introduced into an inert moving gas stream—the *mobile phase* or the *carrier gas*—which carries it into the column containing the *stationary phase*. If the sample is a liquid, then an aliquot of it is heated and its vapor is transferred by the carrier gas into the column. There the sample components are separated, by means of selective interaction (partitioning) between the stationary and the mobile phases. Thus, they emerge at the end of the column at different times and can be detected. The time (*retention time*) that has passed between sample introduction and the emergence of the individual analyte bands—the *peaks*—is, under given conditions, characteristic of the individual analytes, while the size—*height* or *area*—of the individual peaks is proportional to their amount.

It is not our task to discuss the theory and practice of gas chromatography: the reader is referred to the general textbooks (e.g., refs. 1–4). However, from the brief summary just given, one can immediately draw two conclusions as far as the sample and its introduction are concerned. First, it is obvious that sample introduction must be instantaneous: after all, if the sample vapor band introduced into the column already has a significant width, then analyte separation will be hindered by the initial broadness of the analyte mixture's band. Second, it is also obvious that all sample components must be volatile, because otherwise a solid residue will remain in the inlet system. Since this zone is heated, the solid residue may eventually decompose, thus creating volatile breakdown products that get into the column and thus, into the chromatogram, creating the impression that these compounds were present as such in the original sample. Also, sample residue may interfere with subsequent injections, as a result of adsorption and/or catalytic decomposition.

Because of this problem, one may have to follow an indirect procedure in the case of complex solid samples (or samples containing some nonvolatile solid particles), by extracting first the analyte(s) of interest and then introducing an aliquot of the resulting solution into the gas chromatograph. A typical example is the determination of low molecular weight compounds, such as monomers, in a polymer sample. The original traditional methods use solvents either to extract the chemicals of interest or to dissolve the polymer, which is then precipitated. The resulting solution is injected and analyzed by gas chromatography. There are several problems with this approach. First, it is obviously a time-consuming procedure. Second, the analyte will generally be more diluted in the solution than it was in the original sample. And finally, it is difficult to avoid getting polymer into the injector of the gas chromatograph, where any nonvolatile sample constituents will accumulate, causing degradation of the chromatographic performance.

In these cases there is an indirect way in which we can study the volatile compounds present in an essentially nonvolatile matrix, without the necessity of carrying out liquid extraction, or without artificially separating the volatiles from the matrix upon sample handling. That is, we can create a system in which the original (liquid or solid) sample is in a closed container and establish equilibrium conditions between the sample proper and the gas phase above it. In this way, due to the fundamental laws of physics, part of the volatile compounds will be present in the gas phase in contact with the sample proper, and their relative concentrations in the two phases will depend on the partial pressures of these compounds, which, in turn, can be influenced in various ways, among them by the proper selection of temperature. By taking an aliquot of the gas phase, we can analyze the volatile compounds without interference by the nonvolatile matrix. In common practice we refer to the gas phase in contact and in equilibrium with an essentially nonvolatile (or lesser volatile) sample as the *headspace (HS)*, and its investigation as *headspace analysis (HSA)*.

Thus, headspace analysis refers to the analysis of the gas (vapor) phase of a binary heterogeneous system in equilibrium. The other phase may be a liquid or a solid, and we may call it the *condensed phase*. Another expression used is the "analysis of equilibrium vapor"; in German, it is called *Dampfraumanalyse*, vapor volume or vapor phase analysis.

If we carry out extraction with a liquid solvent, the analytes will be distributed between the sample and the solvent, and we adjust the conditions so that their distribution should favor the solvent phase. HSA is also an extraction procedure but now, a gas is used instead of a liquid as the solvent because a gas is an ideal "solvent" for highly volatile compounds. In this case the analytes will be distributed between the condensed phase and the gas phase, and we adjust the conditions to favor the gas phase.

The headspace gas can be investigated by various methods. However, *gas chromatography (GC)* is particularly well suited for such measurements, since GC is an ideal method for gas (vapor) analysis. In *headspace–gas chromatography (HS-GC)*, the vapor (gas) phase in contact with a condensed (liquid or solid) phase is analyzed by gas chromatography.

1.2 Types of Headspace Analysis

HS-GC analysis consists of two steps. First, the sample—a liquid or a solid—is placed in a vessel having a gas volume above it, and the vessel— usually a vial—is closed. This vial is then thermostatted at a constant temperature until equilibrium is reached between the two phases. After equilibrium an aliquot of the vial's gas phase (the *headspace*) is introduced into the carrier gas stream, which carries it into the column, where it is analyzed in the usual way. Figure 1.1 visualizes the two steps of HS-GC. Sample transfer can be carried out in a number of ways: either manually (e.g., by using a gas-tight syringe) or automatically, by means of pressurization of the sample vial and a time- or volume-controlled transfer of an aliquot of the headspace into the column.

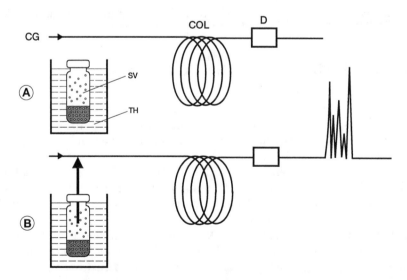

Figure 1.1 Principles of static (equilibrium) headspace–gas chromatography. (A) equilibration and (B) sample transfer. CG = carrier gas; SV = sample vial, TH = thermostat, COL = gas chromatographic column, D = detector.

In this procedure, the two phases in the sample vial are under *static conditions* and sample transfer is carried out after they reach equilibrium. Therefore, we call this type of HS analysis *static* or *equilibrium HS analysis*. This technique is the subject of our book.

Static (equilibrium) headspace analysis is a gas extraction method, and we may further specify it as a *one-step gas extraction*, in which a single aliquot of the gas is analyzed and the information obtained from this one aliquot is used to determine the nature and concentration of the volatile analyte(s) present in the original sample. There is, however, another way to carry out gas extraction. In this version we do not wait for equilibrium and we do not analyze an aliquot of the gas phase: gas extraction is carried out continuously. By continuously removing the gas phase, we rely on the volatile analytes to try to reestablish the equilibrium state, which, however, is never reached. Thus at the end, the total amount of the volatile analytes is removed from the sample. This is *continuous gas extraction*.

In the most frequently used form of continuous gas extraction, an inert gas (the *purge* gas) is passed through the (liquid) sample and the sample solution is extracted by the gas bubbles. The gas effluent is usually conducted through a *trap* containing an adsorbent, which will retard the volatile analytes carried there by the purge gas. When extraction is completed, the collected analytes may be analyzed by rapidly releasing them from the trap (usually by heating and backflushing) and conducting this "slug" into the gas chromatograph. Figure 1.2 visualizes this procedure, which may be called *dynamic headspace analysis* or, more commonly, the *purge-and-trap method*.

In our book, we deal only with equilibrium (static) HS analysis: the techniques of the purge-and-trap method do not represent our subject.

1.3 The Evolution of the HS-GC Methods

"Headspace analysis"—analyzing a gas in contact with a liquid or solid sample and drawing conclusions from the results concerning the nature and/or composition of the original sample—had been carried out well before the development of gas chromatography, or before the combination of the two techniques. Ioffe and Vitenberg, in their book [5], refer to a publication by Harger, Bridwell, and Raney in 1939, in which the possibility of the determination of the alcohol content of aqueous solutions and urine by the analysis of the vapor was indicated [6].

We also know of detailed investigations carried out in 1956–1958 by Schulek and co-workers, in Budapest, investigating the "headspace" of solutions [7–9], and probably other researchers also carried out similar investigations in this period.

The first documented combination of gas chromatography with headspace sampling is the report of Bovijn and co-workers presented at the 1958

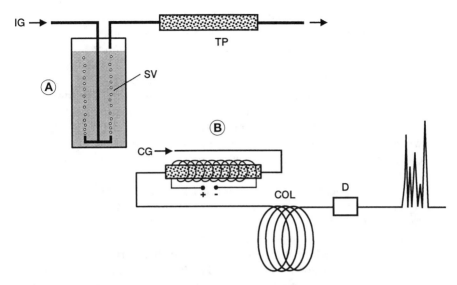

Figure 1.2 Principles of dynamic headspace–gas chromatography ("purge-and-trap"). (A) sample purging and collection of the removed volatiles in a trap and (B) desorption from the trap and transfer into the gas chromatograph. IG = inert purge gas, CG = carrier gas, SV = sample vessel, TP = trap, COL = gas chromatographic column, D = detector.

Amsterdam Symposium [10], on continuous monitoring of the hydrogen content in the water of high pressure power stations. In 1960 W. H. Stahl and co-workers also used headspace sampling for the gas chromatographic analysis of the gas in sealed cans and flexible packages for its oxygen content [11]. We believe this to be the first published use of the expression "headspace." In the case of flexible packages a gas sample was taken directly by a 1-mL hypodermic syringe, while the cans were first pierced by a special device. Around 1962 Beckman Instruments introduced a special Head Space Sampler, intended for the sampling of the headspace of cans or other containers* for their oxygen content. The device had a puncturing tool that was connected to a small closed volume that could be evacuated; in this way, gas was drawn into this sampling volume after piercing the container. The system was designed for use with Beckman's polarographic oxygen sensor, but its brochure [12] indicated the possibility of analyzing the withdrawn gas sample by gas chromatography or spectrophotometry.

With respect to the analysis of volatile organic compounds present in the headspace of foodstuffs, one of us (L.S.E.) utilized HS-GC in the winter of 1958–1959 to follow the development of rancidity of potato chips, by analyzing an aliquot of the headspace gas, taken with a gas-tight syringe directly from the bag of the potato chips. The first published results on the use of (static)

---

* We could not find out whether this device had any connection with the cited work of Stahl.

HS-GC for the investigation of fruits, honey, foods, and aqueous solutions of some organic compounds are from 1961–1962 [13–20].

Static HS-GC was significantly advanced by the pioneering work of Machata who—following the first report of Curry et al. [21] on this possibility—in 1964 described a semiautomatic system for the determination of ethanol in blood [22]. Machata's work opened the way for the development of the first automated instrument to carry out the gas chromatographic analysis of headspace samples (the Perkin-Elmer Model F-40 introduced in 1967 [23, 24]) which, in turn, made it possible to use HS-GC for the routine determination of the ethanol content in blood in the police and forensic laboratories of most countries, replacing the classical Widmark method for this measurement [25, 26]. This application was then followed by many others, eventually making headspace sampling one of the major sample handling procedures in gas chromatography.

The use of continuous gas extraction ("dynamic headspace analysis") was evidently first suggested by Wahlroos [27, 28], but his work was unknown to the general chromatographic community. The first widely known work utilizing such techniques for the analysis of the volatile organic compounds present in breath and urine was described by Teranishi's group [29]. These investigators, however, passed the purge gas over the liquid sample and trapped the removed volatile compounds by freezing. The technique started to be utilized more generally after the introduction of Tenax [poly (2,6-diphenyl-p-phenylene oxide)] as the adsorbent. Tenax was first used in gas chromatography in 1970, by van Wijk [30], but as a column packing: its use as a universal adsorber for dynamic headspace–GC was first proposed by Zlatkis in 1973 [31]. This suggestion was followed by the development of the bubbling-through technique and more detailed investigations on the analysis of volatile urinary metabolites [32, 33], trace volatile organic compounds in blood and plasma [34], and air [35] and water pollutants [36], using this technique.

1.4 Headspace–Gas Chromatography Literature

It is almost impossible to compile a bibliography of the publications dealing with the various HS-GC techniques. The textbooks discussing them in detail [5, 37–39], particularly the book of Ioffe and Vitenberg [5], also give references to the most important publications in this field. Two books have collected the papers presented at three special symposia dealing with HS-GC: 14 papers (mainly on the application of dynamic HS-GC) read at the symposium organized by the Agricultural and Food Chemistry Division of the American Chemical Society, during the 174th National Meeting of the Society (August 29–September 1, 1977, Chicago), were compiled in a collective volume edited by Charalambous [40]. A total of 21 papers presented at the GC-HS symposia held on October 5, 1978, at Beaconsfield, England, and on October

18–20, 1978, at Überlingen, Germany, were compiled by Kolb [41]. We also mention a few very detailed reviews, each containing a large number of literature references. In 1984 Kolb summarized the use of HS-GC for the analysis of food contaminants [42] (70 references); in 1985 McNally and R. L. Grob published a two-part review on environmental and other applications [43, 44] (283 references); in 1990 Namiesnik et al. reviewed the applications of HS-GC related to the analysis of organic compounds in water [45] (495 references); and finally, in 1984, Seto compiled the literature dealing with the determination of volatile substances in biological samples [46] (328 references).

1.5 Regulatory Methods Utilizing (Static) HS-GC

Today static headspace–gas chromatography is widely used in laboratories for various official measurements and for the determination of toxic impurities in the environment. We mention first the determination of ethanol in the blood of drivers: static HS-GC is the accepted official method in many countries. The compilation of regulatory methods for the three countries discussed in this section by no means constitutes a complete listing.

In the *United States*, the Environmental Protection Agency (EPA) has published a number of methods utilizing headspace–gas chromatography, both static and dynamic ("purge and trap"). Static HS-GC is used in the determination of the vinyl chloride content of wastewater and of PVC resin, slurry, wet cake, and latex samples [47–49]. The federal Food and Drug Administration has accepted official methods for the analysis of vinyl chloride in corn oil and food-simulating solvents [50], in oils and vinegar [51], and in PVC food packaging [52], utilizing static HS-GC. The U.S. Pharmacopeia proposes the use of static HS-GC for the analysis of organic volatile impurities (OVIs) [53].

The American Society for Testing and Materials (ASTM) has a number of methods dealing with the use of static HS-GC, mainly in the polymer field [54–59]. Of these we mention particularly the little-known standard method for the determination of residual solvents in flexible packaging materials such as cellophane and polyethylene film [54] which was first accepted as early as 1972. This method already recognized the need for the establishment of the proper equilibrium time and described the progressive working method early in the development of HS-GC.

In *Germany*, DIN and VDI standards* utilizing static HS-GC cover the analysis of water, wastewater, and sludge for benzene and its derivatives [60]; volatile halogenated hydrocarbons [61] and vinyl chloride [62]; measurement of atmospheric pollutants such as vinyl chloride [63] and 1,3-butadiene [64]; and the determination of halogenated hydrocarbons in soil [65]. Static HS-GC was also prescribed by the Committee of the Industrial Association for Food

---

* *DIN*, Deutsche Industrienormen (German Industrial Standards); *VDI* = Verein Deutscher Ingenieure (Association of German Engineers).

Technology and Packaging, to determine the residual solvent content of packaging films [66].

In Germany an official commission deals with the investigation of industrial materials that are detrimental to health, and its Subcommittee on Analytical Chemistry periodically publishes standard analytical methods. The current edition of their compilation [67] lists methods using static HS-GC for the analysis of a wide variety of materials in blood, such as acetone, benzene and alkylbenzenes, dichloromethane, chloroform, carbon tetrachloride, 1,1- and 1,2-dichloroethane, 1,1,2-trichloroethane, 1,2-dichloroethylene, trichloroethylene, tetrachloroethylene, 2-bromo-2-chloro-1,1,1-trifluoromethane (Halothane), 1,4-dioxane, 2-hexanol, isopropylbenzene, carbon disulfide, styrene and organic solvents; as well as 1,1,2-trichloro-1,2,2-trifluoroethane in serum and acetone and organic solvents in urine.

On October 26, 1979, the (West) German Federal Ministry of Health issued specifications limiting the allowable trace monomeric vinyl chloride concentration in consumer products made of PVC and in the food which came in contact with them, indicating static HS-GC as the official analytical method. A similar standard was also adapted on July 8, 1980, by the European Committee for Standardization (CEN) which is an arm of the European Community. Currently CEN is preparing methods for the analysis of other toxic monomers such as acrylonitrile, vinylidene chloride, and vinyl acetate.

In *Japan*, three standards were issued in 1992–1994 to control the allowable concentration of volatile organic compounds present in trace quantities in drinking water and wastewater [68–70]; these documents recommend both static and dynamic HS-GC as analytical methods, and deal also with the mass spectrometric identification of the individual compounds.

This brief summary shows that today, static headspace–gas chromatography is a universally applied analytical method in a wide variety of fields.

References

1. R. L. Grob (editor), *Modern Practice of Gas Chromatography*, 2nd ed., Wiley, New York, 1985; 897 pp.

2. G. Guiochon and C. L. Guillemin, *Quantitative Gas Chromatography for Laboratory and On-Line Process Control*, Elsevier, Amsterdam, 1988; 798 pp.

3. G. Schomburg, *Gas Chromatography: A Practical Course*, VCI Publishers, New York, 1990: 320 pp.

4. L. S. Ettre and J. V. Hinshaw, *Basic Relationships of Gas Chromatography*, Advanstar, Cleveland, 1993; 188 pp.

5. B. V. Ioffe and A. G. Vitenberg, *Headspace Analysis and Related Methods in Gas Chromatography*, Wiley-Interscience, New York, 1984; xv + 276 pp.

6. R. N. Harger, E. G. Bridwell, and B. B. Raney, *Proc. Am. Soc. Biol. Chem., J. Biol. Chem.* **128**, xxxviii (1939); quoted after ref. 5.

7. E. Schulek, E. Pungor, and J. Trompler, *Mikrochim Acta*, **1956** (1/6), 1005–1022.

8. E. Schulek, E. Pungor, and J. Trompler, *Mikrochim Acta*, **1957** (1), 85–95.

9. E. Schulek, E. Pungor, and J. Trompler, *Mikrochim Acta*, **1958** (1), 52–59.

10. L. Bovijn, J. Pirotte, and A. Berger, in D. H. Desty (editor), *Gas Chromatography 1958 (Amsterdam Symposium)*, Butterworths, London, 1958; pp. 310–320.

11. W. H. Stahl, W. A. Voelker, and J. H. Sullivan, *Food Technol.* **14**, 14–16 (1960).

12. *Beckman Head Space Sampler*, Bulletin No. 701, Beckman Instruments, Fullerton, CA, September 1962.

13. C. Weurman, *Food Technol.* **15**, 531–536 (1961).

14. D. A. M. Mackay, D. A. Lang, and M. Berdick, *Anal. Chem.* **33**, 1369–1374 (1961).

15. W. Dörrscheidt and K. Friedrich, *J. Chromatogr.* **7**, 13–18 (1962).

16. S. D. Bailey, M. L. Bazinet, J. L. Driscoll, and A. I. McCarthy, *Food Sci.* **26**, 163–170 (1961).

17. R. G. Buttery and R. Teranishi, *Anal. Chem.* **33**, 1439–1441 (1961).

18. S. D. Bailey, D. G. Mitchell, M. L. Bazinet, and C. Weurman, *J. Food Sci.* **27**, 165–170 (1962).

19. R. Teranishi, R. G. Buttery, and R. E. Lundin, *Anal. Chem.* **34**, 1033–1035 (1962).

20. R. Bassette, S. Özeris, and C. H. Whitnah, *Anal. Chem.* **34**, 1540–1543 (1962).

21. A. S. Curry, G. Hurst, N. R. Kent, and H. Powell, *Nature*, **195**, 603–604 (1962).

22. G. Machata, *Mikrochim Acta*, **1964** (2/4), 262–271.

23. D. Jentzsch, H. Krüger, and G. Lebrecht, *Appl. Gas Chromatogr.* No. 10E, 1967; 21 pp.

24. D. Jentzsch, H. Krüger, G. Lebrecht, G. Dencks, and J. Gut, *Z. Anal. Chem.* **236**, 96–118 (1968).

25. G. Hauck and H. P. Terfloth, *Chromatographia*, **2**, 309–314 (1969).

26. G. Machata, *Blutalkohol*, **4**(5), 3–11 (1967); **7**(5), 345–348 (1970).

27. Ö. Wahlroos, *Ann. Acad. Sci. Fenn. Ser. A. II*, Chemica, **122**, 1 (1963); quoted after ref. 5.

28. Ö. Wahlroos, *Acta Chim. Scand.* **20**, 197 (1966); quoted after ref. 5.

29. R. Teranishi, T. R. Mon, P. Cary, A. B. Robinson, and L. Pauling, *Anal. Chem.* **44**, 18–20 (1972).

30. R. van Wijk, *J. Chromatogr. Sci.* **8**, 418–420 (1970).

31. A. Zlatkis, H. A. Lichtenstein, and A. Tishbee, *Chromatographia*, **6**, 67–70 (1973).

32. A. Zlatkis, H. A. Lichtenstein, A. Tishbee, W. Bertsch, H. Shunbo, and H. M. Liebich, *J. Chromatogr. Sci.* **11**, 299–302 (1973).

33. A. Zlatkis, W. Bertsch, H. A. Lichtenstein, A. Tishbee, F. Shunbo, H. M. Liebich, A. M. Coscia, and N. Fleischer, *Anal. Chem.* **45**, 763–767 (1973).

34. A. Zlatkis, W. Bertsch, D. A. Bafus, and H. M. Liebich, *J. Chromatogr.* **91**, 379–383 (1974).

35. W. Bertsch, R. C. Chang, and A. Zlatkis, *J. Chromatogr. Sci.* **12**, 175–182 (1974).

36. T. A. Bellar and J. J. Lichtenberg, *J. Am. Water Works Assoc.* **66**, 739 (1974).

37. H. Hachenberg and A. P. Schmidt, *Gas Chromatographic Headspace Analysis*, Heyden & Son, London, 1977; xiv + 125 pp.

38. H. Hachenberg, *Die Headspace Gaschromatographie als Analysen- und Messmethode—ein Überblick*, DANI Analysentechnik, Mainz-Kastel, 1988; 52 pp.

39. B. Kolb, *Headspace–Gaschromatographie (HSGC) mit Kapillartrennsäulen*, Vogel Verlag, Würzburg, 1986; 72 pp.

40. G. Charalambous (editor), *Analysis of Foods and Beverages*, Academic Press, New York, 1978; xiv + 394 pp.

41. B. Kolb (editor), *Applied Headspace Gas Chromatography*, Heyden & Son, London, 1980; x + 185 pp.

42. B. Kolb, in J. Gilbert (editor), *Analysis of Food Contaminants*, Elsevier, Amsterdam, 1984; pp. 117–156.

43. M. E. McNally and R. L. Grob, *Am. Lab.* **17**(1), 20–93 (January 1985).

44. M. E. McNally and R. L. Grob, *Am. Lab.* **17**(2), 106–120 (February 1985).

45. J. Namiesnik, T. Górecki, M. Biziuk, and L. Torres, *Anal. Chim. Acta*, **237**, 1–60 (1990).

46. Y. Seto, *J. Chromatogr.* **674**, 25–62 (1994).

47. *EPA Method 107A:* Determination of Vinyl Chloride Content of Solvents, Resin–Solvent Solution, Poly(vinyl chloride) Resin, Resin Slurry, Wet Resin and Latex Samples. *Fed. Regist.* **47**(173) (Sept. 7, 1982).

48. *EPA Method 108A:* Determination of Vinyl Chloride Content of Inprocess Wastewater Samples and Vinyl Chloride Content of Poly(vinyl chloride) Resin, Slurry, Wet Cake and Latex Samples. *Fed. Regis.* **47**(173) (Sept. 7, 1982).

49. *EPA Method D-1-VOA-Q:* Quick Turnaround Method for Contract Laboratory Practice (CLP): Static Headspace Method for Volatile Organic Analytes (VOA) in Soil/Sediments, Employing an Automated Headspace Sampler (November 1989).

50. G. W. Diachenko, C. V. Breder, M. E. Brown, and J. L. Dennison, *J. Assoc. Off. Anal. Chem.* **61**, 570 (1978).

51. B. D. Pace and R. O'Grody, *J. Assoc. Off. Anal. Chem.* **60**, 576 (1977).

52. H. L. Dennison, C. V. Breder, T. McNeal, R. C. Snyder, J. A. Roach, and J. A. Sphon, *J. Assoc. Off. Anal. Chem.* **61**, 813 (1978).

53. *U.S. Pharmacopeia XXIII.* Organic Volatile Impurities (467). Method IV, 1995; pp. 1746–1747.

54. ASTM F-*151-86(91):* Standard Test Method for Residual Solvents in Flexible Barrier Materials.

55. *ASTM D-4526-85:* Standard Practice for Determination of Volatiles in Polymers by Headspace–Gas Chromatography.

56. *ASTM D-3749-87:* Standard Test Method for Residual Vinyl Chloride Monomer in Poly(vinyl chloride) Resins by Gas Chromatographic Headspace Analysis.

57. *ASTM D-4322-83 (91):* Standard Test Method for Residual Acrylonitrile Monomer in Styrene–Acrylonitrile Copolymers and Nitrile Rubber by Headspace–Gas Chromatography.

58. *ASTM D-4443-84(89):* Standard Test Method of Analysis for Determining the Residual Vinyl Chloride Monomer Content in ppb Range in Vinyl Chloride Homo- and Copolymers by Headspace–Gas Chromatography.

59. *ASTM D-4740-93:* Standard Test Method for Residual Vinyl Chloride Monomer in Poly(vinyl chloride) Resins by Gas Chromatographic Headspace Technique.

60. *DIN 38407 (Part 9) (May 1991):* Examination of Water, Wastewater and Sludge: Determination of Benzene and Some of Its Derivatives by Gas Chromatography.

61. *DIN 38407 (Part 5) (November 1991):* Examination of Water, Wastewater and Sludge: Determination of Volatile Halogenated Hydrocarbons by Gas Chromatographic Headspace Analysis.

62. *DIN 38413 (Part 2) (May 1988):* Examination of Water, Wastewater and Sludge: Determination of Vinyl Chloride by Headspace–Gas Chromatography.

63. *VDI Richtlinie 3494 (May 1988):* Measurement of Gaseous Emissions: Determination of Vinyl Chloride Concentration by Gas Chromatography, with Manual or Automatic Headspace Analysis.

64. *VDI Richtlinie 3953 (April 1991):* Measurement of Gaseous Emissions: Determination of 1,3-Butadiene.

65. *VDI Richtlinie 3865 (July 1988):* Measurement of Organic Soil Contaminants: Determination of Volatile Halocarbons in Soil.

66. *Verpack Rundsch.* **40**(7), 56 (1989).

67. D. Henschler (editor), *Analysen in biologischem Material.* Vol. 2: *Analytische Methoden zur Prüfung gesundheitsschädlicher Arbeitsstoffe,* 11th ed., Verlag Chemie, Weinheim, 1994.

68. Japanese Ministerial Ordinance of Drinking Water Quality Standard. Ministry of Health and Welfare, No. 69. *Official Gazette,* December 21, 1992.

69. Japanese Environmental Standard of Water Quality. Notification No. 16 of the Environment Agency. *Official Gazette,* March 8, 1993.

70. Japanese Environmental Standard of Wastewater Quality. Notifications No. 2 and 3 of the Environment Agency. *Official Gazette,* January 10, 1994.

2

Theoretical Background of Headspace–Gas Chromatography and Its Applications

In this chapter we shall discuss the theory of headspace–gas chromatography and the practical application of the theoretical principles.

2.1 Basic Theory of Headspace Analysis

Figure 2.1 visualizes a headspace vial, with the two phases: the *sample* (*condensed*) *phase* and the *gas phase* (*headspace*); we shall indicate them by the respective subscripts of S and G. If the system contains volatile analytes that are soluble in the condensed phase, these will distribute between both phases according to the thermodynamically controlled equilibrium. The system represented by this vial is characterized by the following values:

V_V = total volume of the vial

V_S = volume of the sample phase

V_G = volume of the gas phase

$$V_V = V_S + V_G \tag{2.1}$$

The relative volumes of the two phases in the vial are characterized by the

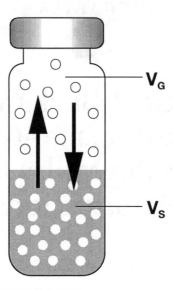

Figure 2.1. A headspace vial containing a liquid sample: V_G = volume of the gas phase, V_S = volume of the liquid sample.

phase ratio β, representing the ratio of the volumes of the two phases present:

$$\beta = V_G/V_S \tag{2.2}$$

$$\beta = \frac{V_V - V_S}{V_S} = \frac{V_G}{V_V - V_G} \tag{2.3}$$

$$V_S = \frac{V_V}{1 + \beta} \tag{2.4}$$

$$V_G = V_V \cdot \frac{\beta}{1 + \beta} \tag{2.5}$$

It is assumed that the volume of the sample phase after equilibrium is equal to the volume of the original sample V_o; in other words, the amount of analyte transferred to the gas phase during equilibration is not considered to result in any appreciable change in the volume of the original sample:

$$V_o = V_S \tag{2.6}$$

The original amount of the analyte in the sample was W_o, and its original concentration was C_o:

$$C_o = W_o/V_S \tag{2.7}$$

After equilibration the respective amounts of the analyte in the two phases are W_S and W_G and their concentrations are C_S and C_G:

$$C_S = W_S/V_S \tag{2.8}$$

$$C_G = W_G/V_G \tag{2.9}$$

$$W_S + W_G = W_o \tag{2.10}$$

The distribution of the analyte between the two phases upon equilibrium is expressed by the thermodynamically controlled *equilibrium constant*. In analogy to the common practice in gas chromatography, the synonymous term *partition (distribution) coefficient* (K) is preferred in this book:

$$K = \frac{C_s}{C_g} \tag{2.11}$$

$$K = \frac{W_S}{V_S} \bigg/ \frac{W_G}{V_G} = \frac{W_S}{W_G} \cdot \frac{V_G}{V_S} = \frac{W_S}{W_G} \cdot \beta \tag{2.12}$$

The partition coefficient is a fundamental parameter that expresses the mass distribution in the two-phase system. It depends on the solubility of the analyte in the condensed phase: compounds with high solubility will have a high concentration in the condensed phase relative to the gas phase ($C_S \gg C_G$); hence, the value of K may be very high. On the other hand, in the case of analytes with little solubility in the condensed phase, C_S will be close to C_G, and might even be less than its value: hence, K will be small.

We can also write the above-derived relationships in the following ways:

$$W_0 = C_o \cdot V_S \tag{2.7a}$$

$$W_S = C_S \cdot V_S \tag{2.8a}$$

$$W_G = C_G \cdot V_G \tag{2.9a}$$

$$C_S = K \cdot C_G \tag{2.11a}$$

Thus, the material balance given by eq. 2.10 can be written as follows:

$$C_o \cdot V_S = C_G \cdot V_G + C_S \cdot V_S = C_G \cdot V_G + K \cdot C_G \cdot V_S = C_G \cdot (K \cdot V_S + V_G) \tag{2.13}$$

Expressing C_o and then C_G:

$$C_o = C_G \left[\frac{K \cdot V_S}{V_S} + \frac{V_G}{V_S} \right] = C_G(K + \beta) \tag{2.14}$$

$$C_G = \frac{C_o}{K + \beta} \tag{2.15}$$

In a given system and under given conditions both K and β are constants, thus $(K + \beta)$ and its reciprocals will also be constants. Therefore, we can write

$$C_G = (\text{const}) \cdot C_o \tag{2.16}$$

In other words, in a given system the concentration in the headspace is proportional to the original sample concentration.

It follows from the basic rules of gas chromatography that the peak area obtained for a given analyte is proportional to the concentration of the analyte in the analyzed sample. In our case, an aliquot of the headspace is analyzed in which the analyte's concentration is C_G. Thus, for the obtained peak area A we can write

$$A = (\text{const.}) \cdot C_G \tag{2.17}$$

where the constant incorporates the influence of a number of analytical parameters and the detector response factor. Combining eqs. 2.16 and 2.17 we can write

$$A = (\text{const}) \cdot C_o \tag{2.18}$$

where the combined constant incorporates the influence of headspace, GC and detector parameters.

We can draw two conclusions from eq. 2.18. The first is that if an aliquot of the headspace at equilibrium is analyzed by GC, the obtained peak area of the analyte will be directly proportional to its concentration in the original sample. This is the fundamental relationship on which quantitative HS-GC analysis is based. The second conclusion is related to the constant in eq. 2.18: as indicated, it incorporates the influence of a number of parameters. Since their numerical evaluation would be very difficult, a prerequisite of reproducible analysis is the exact reproduction of the analytical conditions, and this is particularly true if quantitative measurement is based on the comparative analysis of the sample and a standard.

We can combine eqs. 2.17 and 2.15:

$$A \propto C_G = \frac{C_0}{K + \beta} \tag{2.19}$$

This equation represents the relationship at equilibrium between the peak area A obtained upon analyzing an aliquot of the headspace, the concentration of the analyte in the headspace C_G, the original sample concentration of the analyte C_o, the partition coefficient K, and β, the phase ratio of the vial.

2.2 Basic Physicochemical Relationships

Within a given system (i.e., with a given analyte and solvent) the value of the partition coefficient can be manipulated by changing the analytical conditions. To investigate the factors that influence its value, we use three basic laws: these are Dalton's law, Raoult's law, and Henry's law.

According to *Dalton's law*, the total pressure of a gas mixture is equal to the sum of the partial pressures p_i of the gases present in the mixture:

$$p_{total} = \Sigma p_i \tag{2.20}$$

It follows from Dalton's law that the fraction of the pressure exerted by a gas is equal to the fraction of its total number of moles present in the gas mixture. In other words,

$$\frac{p_i}{p_{total}} = \frac{n_i}{n_{total}} = x_{G(i)} \tag{2.21}$$

$$p_i = p_{total} \cdot x_{G(i)} \tag{2.22}$$

where n represents the number of moles present and $x_{G(i)}$ is the mole fraction of the particular component in the gas mixture. In a dilute gas mixture with a given chemical composition, we can use the concentration of that component in the mixture instead of the mole fraction:

$$x_{G(i)} = (\text{const.}) \cdot C_{G(i)} \tag{2.23}$$

where the constant incorporates the mode of conversion from mole fraction to concentration units. Combining eqs. 2.22 and 2.23 we can write

$$p_i \propto C_{G(i)} \tag{2.24}$$

which means that the concentration of the analyte in the headspace is proportional to its partial pressure.

Raoult's law states that the vapor pressure of a dissolved solute (i.e., its partial pressure p_i) over its solution is directly proportional to its mole fraction in the solution $x_{S(i)}$ and the proportionality constant is the vapor pressure p_i^o of the pure analyte (i.e., when $x_{S(i)} = 1$):

$$p_i = p_i^o \cdot x_{S(i)} \tag{2.25}$$

Raoult's law is valid only for ideal mixtures and in most cases there is a deviation from Raoult's law, as illustrated later (Figures 9.3 and 9.7). To compensate for this deviation, another factor is introduced into eq. 2.25:

$$p_i = p_i^o \cdot \gamma_i \cdot x_{S(i)} \tag{2.26}$$

This parameter is called the *activity coefficient* (γ_i) of compound i. The activity coefficient can be considered as a correction factor to concentration (molality), which modifies it to the true "active concentration" [1].

The activity coefficient depends on the nature of component i and reflects the intermolecular interaction between the analyte and other sample components, particularly the matrix (solvent). Thus the so-called *matrix effect*, which is discussed in detail in Section 4.5, represents the influence of the activity coefficient. For higher concentrations of the analyte, the activity coefficient also becomes a function of the concentration, while in dilute solutions it is constant and independent of the analyte's concentration. In such a so-called *ideal dilute solution* there is a general linearity between the partial vapor pressure and the mole concentration of the analyte, which is expressed by *Henry's law*:

$$p_i = H \cdot x_i \tag{2.27}$$

where H is the Henry's law constant. In an ideal solution ($\gamma_i = 1$), $H = p_i^o$. Henry's law is the basis of gas chromatography and of headspace analysis.

In such an ideal dilute solution (usually $<0.1\%$), each dissolved analyte molecule is surrounded only by solvent molecules. Therefore only the intermolecular interaction forces between solute–solvent molecules are effective, while the probability of additional solute–solute interaction increases with the concentration of the analyte.

Finally, let us substitute p_i from eq. 2.22 into eq. 2.26:

$$p_i = p_{total} \cdot x_{G(i)} = p_i^o \cdot \gamma_i \cdot x_{S(i)} \tag{2.28a}$$

or

$$\frac{x_{S(i)}}{x_{G(i)}} = \frac{p_{total}}{p_i^o \gamma_i} \tag{2.28b}$$

We have already mentioned that in a given system representing ideal dilute solutions, concentration can be substituted for mole fraction (cf. eq. 2.23). Hence:

$$\frac{p_{total}}{p_i^o \cdot \gamma_i} = \frac{x_{S(i)}}{x_{G(i)}} = \frac{C_{S(i)}}{C_{G(i)}} = K \tag{2.29}$$

and therefore

$$K \propto \frac{1}{p_i^o \cdot \gamma_i} \tag{2.30}$$

In other words, the partition coefficient is proportional to the reciprocal of the vapor pressure and the activity coefficient of the analyte: increasing these values will decrease the value of the partition coefficient. On the other hand, according to eq. 2.19, a decrease in the partition coefficient will increase the concentration of the analyte in the headspace at equilibrium, and this results in an increase of headspace sensitivity.

Next, we investigate the ways in which the vapor pressure and the activity coefficient can be changed to favorably influence the partition coefficient. First, let us consider the influence of temperature, the adjustment of which (naturally, with certain limits) is usually under our control.

2.3 Headspace Sensitivity

In headspace analysis we want to obtain a reasonable peak for reliable quantitation when analyzing a given sample. Since the peak area obtained when analyzing an aliquot of the headspace is directly proportional to the concentration of the analyte in the headspace (cf. eq. 2.17), our goal is to achieve a sufficiently high analyte concentration in the headspace upon equilibrium.

The expression "headspace sensitivity" refers to this aim: it is a general expression indicating whether we can achieve a larger or smaller peak area for a given sample concentration. It is obvious that if we can increase the peak area related to a given sample concentration, then we can determine smaller analyte concentrations in the sample.

We have seen that the headspace concentration C_G and thus, the peak area A obtained when analyzing an aliquot of the headspace, can be expressed as follows:

$$A \propto C_G = \frac{C_o}{K + \beta} \tag{2.19}$$

where C_o is the original concentration of the analyte in the sample, K is the distribution (partition) coefficient, and β is the phase ratio relating the volume of the gas phase V_G in the vial (having the volume V_V) to the volume of the condensed phase (sample) V_S:

$$\beta = \frac{V_G}{V_S} = \frac{V_V - V_S}{V_S} = \frac{V_G}{V_V - V_G} \tag{2.2) and (2.3}$$

According to eq. 2.19, headspace sensitivity (the obtained peak area) depends on the *combined* effect of K and β. This is obviously a complex relationship that must be considered from different points of view. For a given original sample concentration of C_o, we can manipulate the headspace concentration C_G of the analyte by changing the values of K and β. In other words, we can change the headspace sensitivity of a system for a given sample by manipulating these two terms. For this reason we must investigate in detail the influence of the partition coefficient K and the phase ratio β (i.e., the sample volume) on the analytical results: the conclusions we shall draw have important consequences for the selection of analytical conditions.

There are two parameters we can easily change: the sample volume in a headspace vial and its temperature. Additional possibilities for changing the whole system include modification of the chemical properties of the sample matrix (e.g., by adding a salt to an aqueous sample) and of the analyte (by preparing more volatile derivatives).

We shall investigate three cases: in the first the temperature of a sample is varied and the influence on compounds with different partition coefficients (K) investigated. In this case the sample volume and thus the phase ratio (β) will be constant and only the partition coefficient is varied, while in the second case the temperature remains constant, with only the sample volume changed to permit evaluation of its influence on compounds with different partition coefficients. A third possibility is then discussed by changing the sample matrix and thus changing the activity coefficient γ_i and thus finally the partition coefficient K. We discuss the possibility of enhancing the headspace sensitivity by producing more volatile derivatives from polar compounds later (see Section 3.9).

Headspace sensitivity may also be enhanced by increasing the volume of the aliquot introduced into the column. Here, the limitations are presented not by the headspace sampling, but by the GC system (column). Techniques permitting the introduction of larger aliquot volumes into the column (e.g., cryofocusing) will be discussed in Chapter 3.

2.3.1 Influence of Temperature on Vapor Pressure and Partition Coefficient

The vapor pressure p_i^o of a pure compound is influenced by its temperature, and the relationship is exponential. It can be described by the following generalized formula:

$$\log p_i^o = -\frac{B}{T} + C \tag{2.31}$$

where B and C are substance-specific constants and T is the absolute temperature.

Figure 2.2 presents this relationship for water; the corresponding values are given in Table 2.1 [2]. As seen, an increase of 20°C in the thermostatting temperature can more than double the vapor pressure. From this it is evident why thermostatting of the sample is important: because of the exponential nature of the relationship in eq. 2.31, even a small temperature change can result in a sizable difference. For example, while for water at 60°C $p_i^o = 19.9\,\text{kPa}$, at 61°C it is $p_i^o = 20.8\,\text{kPa}$: in other words, a one-degree

Table 2.1   Vapor pressure p^o and density d data for saturated water vapor as a function of temperature [2]

°C	p^o (kPa)	p^o (torr) [2a]	d (μg/mL) [2b]
10	1.2	9.2	9.4
20	2.3	17.5	17.3
30	4.2	31.8	30.3
40	7.4	55.3	51.1
50	12.3	92.5	83.2
60	19.9	149.4	130.5
70	31.1	233.7	198.4
80	47.2	355.1	293.8
90	69.9	525.8	424.1
100	101.1	760.0	598.0
110	142.9	1074.5	826.5
120	198.1	1489.1	1122.0

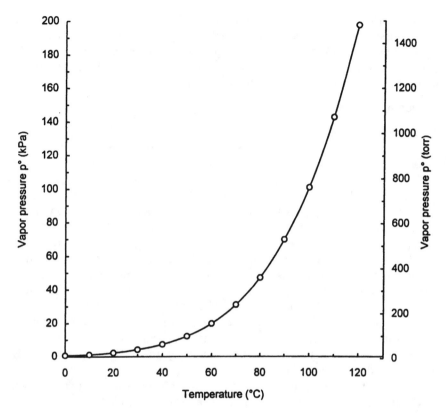

Figure 2.2. Dependence of the saturation vapor pressure p^o of water on temperature. For numerical data see Table 2.1.

difference in temperature resulted in a 4.5% difference in the vapor pressure.

Since the partition coefficient is related to the vapor pressure (eq. 2.30), we can write its relationship with temperature similar to eq. 2.31:

$$K = \frac{(\text{const.})}{p_i^o} \tag{2.32}$$

$$\log K = \log(\text{const}) = \log p_i^o \tag{2.33}$$

Substituting eq. 2.31 for $\log p_i^o$:

$$\log K = \frac{B'}{T} - C' \tag{2.34}$$

where C' now indicates the original constant C from eq. 2.31 plus the constant from eq. 2.32. As seen, eq. 2.34 is similar to eq. 2.31 except that the signs are reversed.

Enhancement of Lower Boiling Compounds. The exponential increase of the vapor pressure with temperature is particularly apparent in the case of a homologous series. Within one homologous series there exists a relationship between the vapor pressure (p_i^o) and the number of carbon atoms ("carbon number," c_n) in the molecules of homologues [3]:

$$\log p_i^o = -a \cdot c_n + b \tag{2.35}$$

where a and b are constants.

Because of this relationship, if we have a mixture of n-paraffins (or other homologues) as the sample in which each component is present in equal concentration, the partial vapor pressures, hence the headspace concentrations of the lower boiling components (components with smaller carbon number) will be much higher than those of the higher homologues.* This follows from eq. 2.26, which described the relationship between the vapor pressure and the partial pressure, and from eq. 2.24, which related the partial pressure of the analyte to its concentration in the headspace. Thus we can conclude that much smaller sample concentrations of the lower boiling components can be determined in a sample by analyzing the headspace than by means of the direct analysis of the original (liquid) sample. This rule is illustrated by Figure 2.3, which compares the chromatogram obtained when analyzing the original sample with that of the headspace. While the peaks for n-alkanes up to dodecane cannot even be seen in the chromatogram of the original (liquid) sample, they dominate the headspace chromatogram; on the other hand, while n-heneicosane ($n\text{-}C_{21}H_{44}$) is the highest concentration of the original sample, its peak in the headspace chromatogram is almost negligible [4].

2.3.2 Influence of Temperature on Headspace Sensitivity for Compounds with Differing Partition Coefficients

Table 2.2 lists the partition coefficient values of a number of analytes for air–water systems [5, 6], together with the regression constants B' and C' of eq. 2.34 and the correlation coefficient r. As seen, increasing the temperature reduces the value of the partition coefficient, which is inversely proportional to the increase in the vapor pressure according to eqs. 2.30 and 2.32. Since, however, the resulting headspace sensitivity depends not only on the partition coefficient but also on the phase ratio, we have to investigate the influence of both parameters.

The partition coefficient of a given analyte can be changed by changing the temperature. We saw the relationship fo K versus T earlier:

$$\log K = \frac{B'}{T} - C' \tag{2.34}$$

---

* It is assumed here that changes in the activity coefficient of the individual homologues do not compensate for the differing partial vapor pressures.

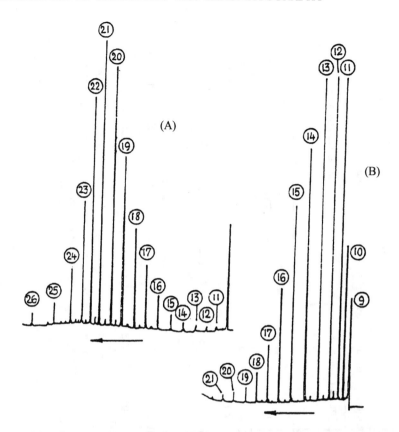

Figure 2.3. Chromatograms obtained when analyzing a paraffin mixture. (A) Analysis of the original liquid sample. (B) Analysis of the headspace in equilibrium at 80°C with the liquid sample. *GC conditions:* Column 50 m × 0.25 mm I.D. open-tubular, coated with OV-17 phenyl (50%) methyl silicone oil. Column temperature: isothermal at 80°C for 2 minutes, then programmed at 7°C/min to 210°C. Split ratio: 1/30. Flame-ionization detector. *Peaks:* the numbering of the peaks indicate the carbon numbers of the individual *n*-paraffins.

Source: Reproduced from ref. 4 with the permission of *Journal of Chromatographic Science.*

Thus, an increase in temperature decreases the value of the partition coefficient. However, the actual influence on headspace sensitivity of this change depends on the relative value of K versus the phase ratio β according to eq. 2.19. Headspace sensitivity (the obtained peak area) depends on the *combined* effect of K and β. Equation 2.19 can be written in the following form:

$$A_G \propto C_G = \alpha \cdot C_o \tag{2.19a}$$

$$\alpha = \frac{1}{K + \beta} \tag{2.36}$$

Table 2.2 Partition coefficient values of selected compounds in air–water systems, at different temperatures*

Compound	Partition coefficient K at					Linear regression data		
	40°C	50°C	60°C	70°C	80°C	B'	C'	r
Dioxane	1618	(1002)	642 (641)	412	288 (283)	2086.8	3.456	0.99977
Ethanol	1355	(820)	511 (512)	328	216 (216)	2205.3	3.910	0.99999
Isopropyl alcohol	825	(479)	286 (290)	179	117 (115)	2351.7	4.597	0.99988
n-Butanol	647	(384)	238 (236)	149	99 (96.7)	2277.5	4.463	0.99955
Methyl ethyl ketone	139.5[†]	(109)	68.8 (72.0)	47.7	35.0 (33.7)	1936.3	3.955	0.99774
Ethyl acetate	62.4	(42.7)	29.3 (30.6)	21.8	17.5 (16.7)	1548.5	3.161	0.99691
n-Butyl acetate	31.4	(20.6)	13.6 (14.3)	9.82	7.58 (7.25)	1726.3	4.028	0.99758
Benzene	2.90[†]	(3.18)	2.27 (2.20)	1.71	1.66 (1.58)	836.2	2.168	0.97544
Toluene	2.82	(2.23)	1.77 (1.82)	1.49	1.27 (1.25)	963.1	2.631	0.99822
o-Xylene	2.44	(1.79)	1.31 (1.40)	1.01	0.99 (0.89)	1148.3	3.302	0.97481
Dichloromethane	5.65	(4.29)	3.31 (3.32)	2.60	2.07 (2.07)	1205.9	3.099	0.99999
1,1,1-Trichloromethane	1.65	(1.53)	1.47 (1.40)	1.26	1.18 (1.19)	413.4	1.095	0.97582
Tetrachloroethylene	1.48	(1.28)	1.27 (1.09)	0.78	0.87 (0.82)	749.1	2.210	0.88046
n-Hexane	0.14	(0.068)	0.043 (0.031)	0.012	(0.0075)	3634.5	12.416	0.97255
Cyclohexane	0.077	(0.055)	0.040 (0.040)	0.030	0.023 (0.23)	1453.7	5.758	0.99990

* The listed partition coefficients represent measured values [5, 6]. The values of B', C' and the correlation coefficient r were calculated by linear regression of measured values according to eq. 2.34. The partition coefficient values in parentheses were calculated from the linear regression data, according to eq. 2.34.
† At 45°C.

If we investigate the temperature influence on the volatility of a certain compound in a sample, we assume a constant value of C_o in such comparative measurements. In this case,

$$A_G \propto \alpha \tag{2.19b}$$

Thus, in the case of a constant C_o, headspace sensitivity is directly related to changes in α: as a result of the inverse relationship expressed in eq. 2.36, a higher value of α means a smaller value for $(K + \beta)$. However, it should be noted that the values of K for different compounds may vary over four orders of magnitude (cf. Table 2.2), while the phase ratio β is usually not changed as much—for example, if we increase the sample volume in a 20 mL vial from 1 mL up to 10 mL we change β from 19 to 1 only. We therefore can consider the following cases:

- If K is much greater than β, then

$$\alpha = \frac{1}{K + \beta} \to \frac{1}{K} \tag{2.37a}$$

and the headspace sensitivity depends directly on K: the higher the temperature, the higher the headspace sensitivity. In this case the phase ratio (i.e., the sample volume) has practically no influence, or only very little. Since K changes significantly with temperature, the effect of temperature on headspace sensitivity is considerable.
- However, if K is smaller than β, then the value of α is primarily influenced by β:

$$\alpha = \frac{1}{K + \beta} \to \frac{1}{\beta} \tag{2.37b}$$

and the headspace sensitivity is determined by the value of the phase ratio: the partition coefficient has no, or only little influence. On the other hand, the phase ratio does not depend on temperature: this means that in such a case, the thermostatting temperature has almost no influence on headspace sensitivity.

To investigate this temperature effect we used an aqueous solution of five solutes with a wide range of partition coefficients: these are ethanol, methyl ethyl ketone, toluene, tetrachloroethylene, and n-hexane. Their partition coefficient values at different temperatures are listed in Table 2.2. The sample volume was 5 mL in a 22.3 mL vial (phase ratio $\beta = 3.46$) and the concentrations were adjusted to yield comparable peak area values at the beginning temperature of 40°C. The original concentration C_o of each compound remained the same for each measurement at increasing sample temperature.

The foregoing conclusions are confirmed and illustrated by Figure 2.4, which plots the resulting peak area values of these five compounds versus increasing sample temperatures. It is obvious that the increase of the tempera-

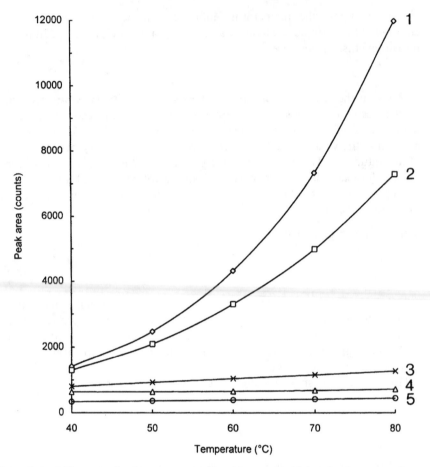

Figure 2.4. Influence of temperature on headspace sensitivity (peak area values, counts) as a function of the partition coefficient K from an aqueous solution with $\beta = 3.46$. Table 2.2 lists the numerical data of K for ethanol (1), methyl ethyl ketone (2), toluene (3), n-hexane (4), and tetrachloroethylene (5).

ture enhances the headspace sensitivity for the polar compounds with high partition coefficients only, while the nonpolar compounds with low K are practically not affected. This is easy to understand, because the temperature affects the fraction of the analyte in the condensed phase only and this fraction is determined by both the partition coefficient K and the phase ratio β: the higher this fraction, the stronger the temperature effect. The following conclusions therefore are immediately apparent from Figure 2.4:

• In the case of ethanol and methyl ethyl ketone, we are in the region where $K \gg \beta$. Thus, the headspace sensitivity depends directly on K and is practically proportional to the change in the partition coefficient.

- In the case of toluene and tetrachloroethylene the partition coefficient is already smaller than the phase ratio. Here, change in the temperature (i.e., in the partition coefficient) has very little influence on the headspace sensitivity.
- In the case of n-hexane, $K \ll \beta$. Now, changes in the partition coefficient due to changes in temperature have no influence on headspace sensitivity

These examples again demonstrate that the influence of temperature is an analyte-specific function and must be evaluated separately for each case.

It follows that in each practical case the temperature for a given sample must be selected according to a number of criteria. One should always try to select the lowest temperature possible, since certain additional limitations must be considered:

- Some samples may be sensitive to higher temperatures and may either be decomposed or oxidized by the air in the headspace vial.
- Let us not forget that the pressure in the headspace is the sum of all partial pressures (cf. eq. 2.20). In the case of solutions, the solvent's vapor mainly determines the pressure in the headspace, while the concentrations of the dissolved solutes are usually small enough to permit the analyst to neglect their contribution to the overall pressure. Thus, particularly when the solvent (the matrix) has a relatively low boiling point, it can significantly increase the headspace pressure. It is obvious that if organic solvents are required, those with higher boiling points should be preferred (e.g., a glycol ether replacing ethanol). Table 4.2 in Chapter 4 lists such solvents.
- If the pressure in the headspace vial is too high, some instrumental problems may result. For example, the removal of an aliquot by a syringe (either manually or using an automated system) will become difficult. As we shall see in Chapter 3, the operation of most automatic headspace analyzers is based on pressurization of the sample vial, followed by the transfer of an aliquot of the headspace to the column or to a loop. An additional pressurization needs an even higher pressure, and such excessive pressures in the vial may cause leakage or even breakage.

2.3.3 Influence of Sample Volume on Headspace Sensitivity for Compounds with Differing Partition Coefficients

In the preceding examples, the phase ratio (hence, the sample volume) was constant. Let us now investigate how changes in the sample volume affect headspace sensitivity for compounds with different partition coefficients. Again we consider the parameter α:

$$\alpha = \frac{1}{K + \beta} \tag{2.36}$$

For convenience we express the relative sample size in the vial in these investigations as the *phase fraction* (ϕ_S) of the sample [7]. This is a term similar to the phase ratio, except that it now gives the sample volume V_S as a fraction of the total volume of the vial V_V:

$$\phi_S = V_S/V_V \tag{2.38}$$

The phase fraction ϕ_S and the phase ratio β are interrelated terms:

$$\phi_S = \frac{1}{1+\beta} \tag{2.39a}$$

$$\beta = \frac{1-\phi_S}{\phi_S} \tag{2.39b}$$

The phase fraction is an immediately obvious term: for example, if a vial is filled up to 50% of its volume with the sample, ϕ_S will be 0.5. We shall consider two values for ϕ_S, 0.2 and 0.8, representing 20 and 80% of the vial's volume (corresponding to the respective phase ratio values of 4.0 and 0.25); sample volumes much smaller than 20% of the vial could be used, and in fact, many times only a few microliters of the sample will suffice*; however, 80% is about the upper limit. We should not forget that there must be enough headspace volume available for transfer into the column. For packed columns and larger-diameter, open-tubular columns, or for conventional open-tubular columns used in the split mode, typical gas volumes transferred into the column are as high as 0.5–2.0 mL. One also needs enough space to ensure that the needle of the headspace sampling system will not protrude into the actual (liquid) sample. For each phase-ratio value, we shall investigate the situation with four values of the partition coefficient: 0.2, 1.0, 20, and 250. Table 2.3 gives numerical values of α for the eight cases:

- If $K = 0.20$, then when filling the vial to 80% of its volume by the sample, the headspace sensitivity (the concentration of the analyte in the headspace) will be almost 10 times that when the sample occupies only 20% of the vial's volume.
- At $K = 1.0$, the headspace sensitivity still changes significantly with changes in the sample volume: there is a fourfold difference between 20 and 80% filling of the sample vial.
- At $K = 20$, the change in headspace sensitivity is more modest, only 18.5% for the given range.
- Finally, at $K = 250$, a fourfold change in the sample volume has almost no influence on headspace sensitivity: the change is only 1.5%.

---

* Such small sample volumes are used in general only with the total vaporization technique (TVT), aiming at the preparation of a vapor standard in which only a single phase is present (see Section 4.6.1).

Table 2.3 Change in headspace sensitivity upon changing the sample volume*

K	ϕ_g	β	$\alpha^\dagger$	Difference in α
0.20	0.20	4.00	0.238	
	0.80	0.250	2.222	× 0.34
1.00	0.20	4.00	0.200	
	0.80	0.250	0.800	× 4
20	0.20	4.000	0.04167	
	0.80	0.250	0.04938	+ 18.5%
250	0.20	4.000	0.00394	
	0.80	0.250	0.00400	+ 1.5%

* Vial volume: 22.3 mL.
$\dagger$ $\alpha = 1/(K + \beta)$.

Table 2.2 lists the partition coefficients of a number of compounds at different temperatures in air–water systems. As seen, dioxane, alcohols, esters, and ketones have high K values; on the other hand, paraffinic, aromatic, and halogenated hydrocarbons have K values in the single-digit range, or even below unity. Thus with such compounds, the sample volume has a crucial role. This is illustrated in Figure 2.5, which shows the analysis of an aqueous solution of cyclohexane and dioxane, thermostatted at 60°C, using 1.0 and 5.0 mL sample volumes; the corresponding peak area values are listed in Table 2.4.

In the case of dioxane ($K = 642$ at 60°C), the peak area changed only by 1.3% (71,848 vs. 72,800); on the other hand, in the case of cyclohexane ($K = 0.040$ at 60°C), a 5.5-fold change (42,882 vs. 237,137) was obtained for the fivefold change in the sample volume.

Sample-to-Sample Reproducibility. In routine analysis, since we need to be able to judge the importance of the exact reproducibility of the sample volume, we must know the influence of the sample volume relative to the value of the partition coefficient. To investigate this question, let us consider the analysis in triplicate of a sample having a mean volume of 10 mL ($V_V = 22.3$ mL,

Table 2.4 Peak area values for the chromatograms shown in Figure 2.5*

	Peak area (counts)	
Sample	Cyclohexane	Dioxane
A: 1.0 mL solution	42,882	71,848
B: 5.0 mL solution	237,137	72,800
C: 5.0 mL solution + 2 g NaCl	240,287	234,312

* For details, see the caption of Figure 2.5.

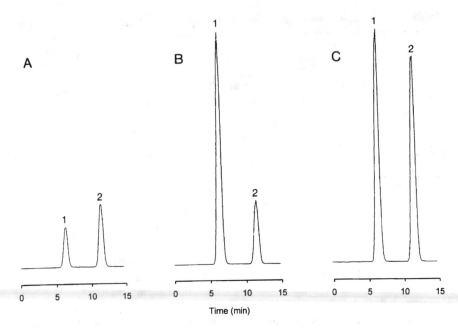

Figure 2.5. Analysis of three samples of an aqueous solution of cyclohexane (0.002 vol %) and 1,4-dioxane (0.1 vol %) in a 22.3 mL vial: (A) 1.0 mL solution ($\beta = 21.3$); (B) 5.0 mL solution ($\beta = 3.46$); (C) 5.0 mL solution ($\beta = 3.46$) to which 2 g NaCl was added. *HS conditions:* equilibration at 60°C, with shaker. *Peaks:* 1 = cyclohexane, 2 = 1,4-dioxane.

$\phi_S = 0.448$, $\beta = 1.23$), with a ± 1 mL change in the first and third samples, representing a $\pm 10\%$ variation. Considering four analytes with the partition coefficient values given earlier, we can investigate the influence of this ± 1 mL change in sample volume on headspace sensitivity. Numerical results are given in Table 2.5. We can conclude that sample volume reproducibility is critical only for low partition coefficient values; in the case of higher values, its influence on the reproducibility of analytical results is negligible. Since, however, the value of the partition coefficient is generally not known, it is always advisable to reproduce the sample volume in duplicate analyses as closely as possible.

2.3.4 Changing the Sample Matrix by Varying the Activity Coefficient

The activity coefficient γ describes the intermolecular interaction between solute and solvent. It is thus a property of both the analyte and the sample matrix. By changing the sample matrix, we can modify the activity coefficient and finally the partition coefficient and the headspace sensitivity.

Table 2.5 Influence of the variation of sample volume on headspace sensitivity*

K	V_S (mL)	ϕ_S	β	$\alpha^\dagger$	Variation from medium value (%)
0.2	9	0.404	1.478	0.597	−14.5
	10	0.448	1.230	0.698	
	11	0.493	1.027	0.814	+16.6
1.0	9	0.404	1.478	0.404	−9.8
	10	0.448	1.230	0.448	
	11	0.493	1.027	0.493	+10.0
20	9	0.404	1.478	0.0466	−1.06
	10	0.448	1.230	0.0471	
	11	0.493	1.027	0.0476	+1.06
250	9	0.404	1.478	0.00398	−0.075
	10	0.448	1.230	0.00398	
	11	0.493	1.027	0.00398	+0.10

* Vial volume: 22.3 mL.
† $\alpha = 1/(K + \beta)$.

The partition coefficient is inversely proportional to both the vapor pressure p_i^o and the activity coefficient γ_i of the analyte:

$$K \propto \frac{1}{p_i^o \cdot \gamma_i} \tag{2.30}$$

As already discussed, the activity coefficient generally differs from unity; however, we can assume in most cases that it is constant under given conditions.

According to eq. 2.30, the value of the partition coefficient could also be decreased by increasing the value of the activity coefficient. A smaller partition coefficient indicates reduced solubility of the analyte in the matrix, and thus an increased concentration in the headspace (eq. 2.11); in turn this increases headspace sensitivity (cf. eq. 2.19). In the case of aqueous solutions of polar compounds, this can be achieved by adding an electrolyte to the sample. This technique, commonly called *salting out*, has been in use in analytical and preparative chemistry for a long time. Figure 2.6 presents an illustrative analysis of volatile halogenated hydrocarbons in water, with and without the addition of an inorganic salt. Clearly the effect is not the same for every analyte, since the more polar compounds with an active hydrogen (dichloromethane and the haloforms) are more affected than the nonpolar compounds (e.g., carbon tetrachloride). However, the sensitivity is not influenced by the salting-- out effect alone; there is an additional volume effect involved which is widely ignored. The addition of a high amount of salt increases the volume of the liquid sample and thus decreases the phase ratio β. For example, the addition of 6 g of potassium carbonate to a 5 mL sample increases the liquid volume

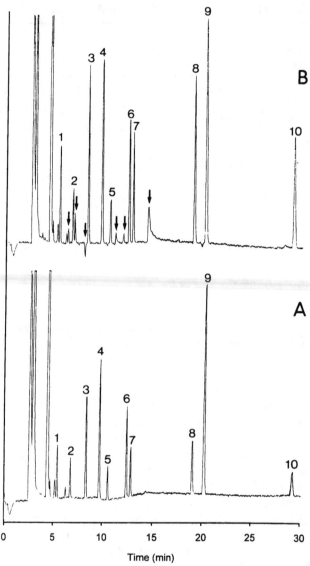

Figure 2.6. The effect of salt addition to an aqueous solution containing low ppb concentrations of volatile halogenated hydrocarbons: analysis of (A) a 5 mL sample and (B) a 5 mL sample to which 6 g K$_2$CO$_3$ was added. *Headspace conditions:* thermostatting at 60°C for 1 hour, with shaker. *GC conditions:* Column: 50 m × 0.32 mm I.D. open-tubular, coated with SE-54 phenyl (5%) vinyl (1%) methyl silicone oil; film thickness: 2 μm. Electron capture detector. Carrier gas: Helium, 150 kPa; makeup gas: argon–methane. Splitless sampling. *Peaks:* 1 = dichloromethane; 2 = 1,1-dichloroethane, 3 = chloroform, 4 = 1,1,1-trichloroethane, 5 = carbon tetrachloride, 6 = trichloroethylene, 7 = dichlorobromomethane, 8 = dibromochloromethane, 9 = tetrachloroethylene, 10 = bromoform. Peaks marked with an arrow are impurities from the salt.

up to 6.5 mL and changes β from 3.46 to 2.43. Already this volume effect seems to increase the sensitivity according to eq. 2.19, particularly for samples with a low partition coefficient K, as already discussed in Section 2.3.3. For these samples it would be better to achieve a smaller phase ratio β and thus a higher sensitivity by increasing the volume of the aqueous sample rather than by adding a salt. Strong salt concentrations increase the viscosity of water samples and therefore prolong the necessary thermostatting time. Moreover, salt often contains volatile impurities, as can be seen in Figure 2.6B, although in this case the potassium carbonate had been cleaned by heating at 250°C overnight.

To be effective, high salt concentrations are needed. In Section 4.5, discussing matrix effects, we show that in the case of an aqueous solution of halogenated hydrocarbons at the ppb level, salt concentrations up to 5% have practically no influence on the peak area obtained by analyzing an aliquot of the headspace under identical conditions. As a comparison, the addition of 20% Na_2SO_4 to a 260 ppb solution of trichloroethylene, increased the peak area by 19.3%.

Figure 2.5 (B vs. C) gives another example for the effect of salt addition. Here, 2 g NaCl was added to a 5.0 mL aqueous solution of dioxane (0.1 vol %) and cyclohexane (0.002 vol %). As indicated by the data in Table 2.4, the addition of such a significant amount of an inorganic salt increased the peak area for the polar dioxane by a factor of 3.22; the small difference of 1.3% for cyclohexane with a K value of practically zero was probably caused by the aforementioned volume effect.

As a conclusion it can be stated that salting out may be a way to increase headspace sensitivity: however, it influences only the distribution of polar compounds, and it is effective only when high salt concentrations are used. In addition, it is an empirical technique: one cannot predict the relative influence of salt addition on the various analytes. Furthermore, salting out influences the linear range of various analytes. Therefore, the technique must be used carefully, and its effect investigated, before it is adapted for quantitative routine analysis.

A similar effect can also be obtained by the addition of water to a solution in an organic solvent that is miscible with water. Many organic compounds have a higher solubility in the organic solvent than in water: therefore their partition coefficient is higher in the organic solvent, which means that their headspace sensitivity will be poorer. Upon addition of water the situation changes: solubility—hence the partition coefficient—is reduced; and thus, headspace sensitivity is increased. Naturally, however, one must be careful that the components of interest are still dissolved in the solvent–water mixture.

This effect was first utilized by Steichen [8] for the analysis of residual 2-ethylhexyl acrylate (EHA) in a polymer. The sample was first dissolved in dimethylacetamide (DMA), and then water was added to the solution. By adding 5 mL water to 2 mL of the DMA solution, a 600-fold increase in the headspace sensitivity was achieved for EHA.

A similar example was presented by Hachenberg and Schmidt [9a], demonstrating the increase of the headspace sensitivity for C_2–C_5 alcohols dissolved in dimethylformamide (DMF) upon the addition of water. Their data are presented in Table 2.6 and Figure 2.7. Considering, for example, the case of a solvent consisting of a 40:60 mixture of water and DMF, a 1.3- to 6.2-fold increase was obtained for the peak area of the individual sample components relative to the use of pure DMF.

This example also shows another interesting effect: namely, that it is possible to change the relative sensitivity for different compounds by changing the composition of the solvent mixture. In pure DMF, ethanol has the highest and n-pentanol the lowest peak area for equal concentration. This indicates that ethanol has a relatively lower solubility in DMF as compared to n-pentanol. On the other hand, the situation is reversed in water in which ethanol has a higher solubility: that is a higher partition coefficient than pentanol, and therefore, a relatively lower headspace sensitivity. Around a solvent concentration of 70:30 water/DMF, the headspace sensitivity for all homologous alcohols is equal, since at this point changes in the activity coefficients compensate for changes in the vapor pressure, so that the product $p_i^0 \cdot \gamma_i$ is constant for the investigated analytes. We can also see that the overall headspace sensitivity for the alcohols is higher in water than in DMF: for each analyte, a higher peak area was obtained in water than in DMF.

Ioffe and Vitenberg [10] treated this situation as a case of the additivity of the partition coefficients in mixed solvents: according to them, in an ideal case (both γ_1 and γ_2 are equal to unity) the relationship should be

$$\log K_{1,2} = x_1 \log K_1 + x_2 \log K_2 \tag{2.40}$$

where K_1 and K_2 are the partition coefficients of the analyte in pure solvents, and $K_{1,2}$ is its partition coefficient in the mixed solvent; x_1 and x_2 are the mole fractions of the two solvents. However, deviations due to the nonideal case (i.e.,

Table 2.6 Headspace sensitivity for C_2–C_5 alcohols as a function of the water content of the solvent*

Solvent composition (%)		Peak area (arbitrary units)			
DMF	Water	Ethanol	n-Propanol	n-Butanol	n-Pentanol
100		12.1	4.9	2.1	0.9
80	20	13.3	7.3	4.2	2.4
60	40	15.7	11.4	8.5	5.6
40	60	18.9	16.6	14.9	14.2
20	80	22.1	23.3	25.7	29.7
	100	26.0	32.5	44.5	63.2

* Concentration: 120 ppm of each analyte in the solution. The peak area values were corrected for variation in detector response for the individual compounds.
Source: Ref. 9a.

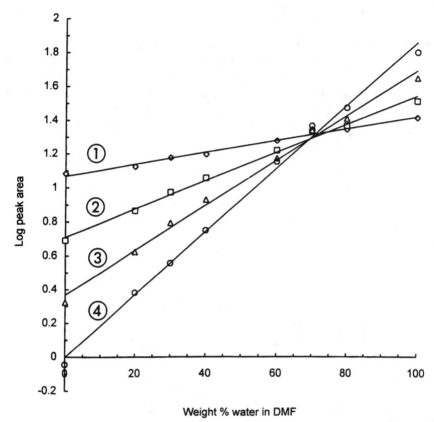

Figure 2.7. Headspace sensitivity for C_2–C_5 alcohols in solutions consisting of dimethylformamide and water, in various ratios. Concentration of each analyte in the solutions: 120 ppm. The measurements were carried out under identical conditions. The peak area values, in arbitrary units, were corrected for variations in the detector response for the individual analytes: $1 =$ ethanol, $2 = n$-propanol, $3 = n$-butanol, $4 = n$-pentanol.

Source: After Hachenberg and Schmidt [9a].

$\gamma \neq 1$) can be significant: in this respect let us not forget the meaning of the activity coefficient as a correction factor to concentration, modifying it to a true "active concentration." Correspondingly, in nonideal cases, eq. 2.40 would be modified to

$$\log K_{1,2} = \gamma_1 x_1 \log K_1 + \gamma_2 x_2 \log K_2 \tag{2.41}$$

where γ_1 and γ_2 are the respective activity coefficients.

Headspace–gas chromatography can be used to study the variation of the activity coefficients [9b, 11, 12], but this is discussed in Chapter 9.

2.4 Headspace Linearity

The theoretical treatment of HS-GC assumes that in the investigated range, both the partition coefficient and the activity coefficient are constant and independent of analyte concentration. Thus, "headspace linearity" refers to the linear relationship between the original concentration of the analyte in the sample (C_o) and its concentration in the headspace (C_G) or between C_o and A, the peak area obtained when analyzing an aliquot of the headspace.

   The actual linear range depends on the analyte's solubility (i.e., its partition coefficient) and its activity coefficient. Generally it covers concentrations below 0.1–1%, which is usually the range being analyzed by HS-GC. Figure 2.8 shows linearity plots for three halocarbons using a flame-ionization detector. In these plots the smallest concentrations measured were 0.1–0.5 ppm, but the

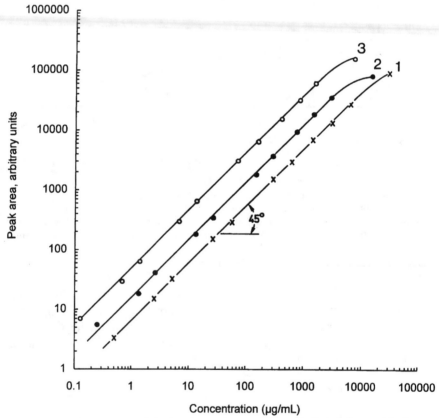

Figure 2.8. Linear range of the headspace response for volatile halogenated hydrocarbons in water, at 80°C. Sample volume: 1 mL. Flame-ionization detector. *Peaks:* 1 = chloroform, 2 = trichloroethylene, 3 = carbon tetrachloride.

linear range could be extended to smaller values by using an electron-capture detector and/or higher sample volumes. At 80°C volatile halogenated hydrocarbons have partition coefficient values around 1–2, where an increase of the sample volume (phase ratio) will increase headspace sensitivity and thus improve minimum detectability.

In some cases the linear range extends to higher concentrations. A typical case is the analysis of aqueous ethanol solutions (Figure 2.9), where the linear range extends up to 25–30%. This observation permits the practical conclusion that the direct determination of the ethanol content of blood or wines is possible by HS-GC. On the other hand, in the case of hard liquor (brandies, gin, vodka, whiskies) the sample must first be diluted.

The actual linear range of HS-GC for a given analyte cannot be predicted: it must be determined by actual measurements. These can be carried out automatically by the multiple headspace extraction (MHE) technique, described shortly (Section 2.6), while a practical example for this particular application is shown in the last chapter (cf. Figure 9.10).

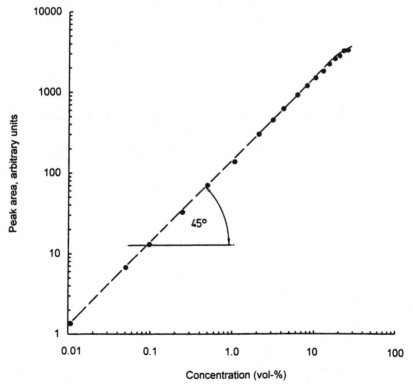

Figure 2.9. Linear range of the headspace response for ethanol in water at 60°C. Sample volume: 1 mL. Flame-ionization detector.

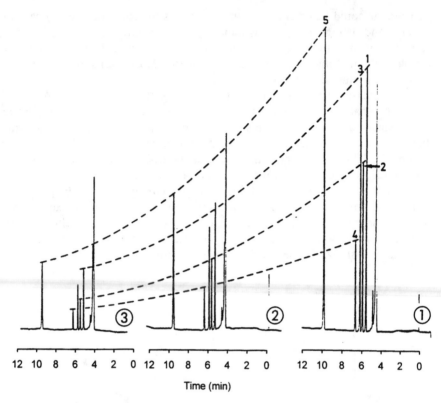

Figure 2.10. Three consecutive MHE analyses of an aqueous solution of volatile halogenated hydrocarbons. *HS conditions:* 1 mL sample, equilibrated at 80°C for 30 minutes. *GC conditions:* 50 m × 0.25 mm I.D. fused-silica, open-tubular column coated with bonded methyl silicone stationary phase; film thickness: 1 μm. Column temperature: 45°C. Electron capture detector. *Peaks* (concentration): 1 = chloroform (25 μg/L), 2 = 1,1,1-trichloroethane (5 μg/L), 3 = carbon tetrachloride (0.5 μg/L), 4 = trichloroethylene (4 μg/L), 5 = tetrachloroethylene (2 μg/L).

2.5 Duplicate Analyses

In general GC analysis, duplicate analysis means taking two consecutive aliquots from the same sample. For example, when an autosampler is used in which the (liquid) sample is in a vial closed with a septum, the instrument will withdraw an aliquot of the sample with the automated syringe and inject it into the gas chromatograph. When the analysis is finished, the instrument withdraws another aliquot of the same volume from the same sample vial and reanalyzes it under identical conditions. Theoretically, the two analyses should give the same result (peak area). Thus, the deviation of the two measurements will indicate the precision of the GC analyses.

If we were to carry out the same measurement in HS-GC, the two results may be different, for the following reason. As we have seen, under equilibrium conditions we have the respective concentrations of C_S and C_G in the sample and gas phase with volumes V_S and V_G; the respective amounts of the analyte in the two phases are W_S and W_G, and the equilibrium is characterized by the partition coefficient K:

$$C_{S,1} = W_{S,1}/V_{S,1} \tag{2.42}$$

$$C_{G,1} = W_{G,1}/V_{G,1} \tag{2.43}$$

$$W_{S,1} + W_{G,1} = W_{o,1} \tag{2.44}$$

$$K = \frac{C_{S,1}}{C_{G,1}} = \frac{W_{S,1}}{W_{G,1}} \cdot \beta \tag{2.45}$$

where subscript 1 indicates the first of two consecutive measurements.

We now withdraw from the headspace an aliquot that contains W_A of the analyte, and then allow the vial to reequilibrate. Because we withdraw a part of the analyte, its total amount in the vial is now less than before the sample withdrawal: it is now $(W_o - W_A)$. At the new equilibrium, we shall have the respective amounts of $W_{S,2}$ and $W_{G,2}$ present in the two phases, with the corresponding new concentrations of $C_{S,2}$ and $C_{G,2}$; however, the distribution coefficient will remain the same. We can thus write the following relationship for this second equilibration, after the withdrawal of an aliquot of the original headspace:

$$C_{S,2} = W_{S,2}/V_S \tag{2.46}$$

$$C_{G,2} = W_{G,2}/V_G \tag{2.47}$$

$$W_{G,2} + W_{S,2} = W_{o,2} = W_{o,1} - W_A \tag{2.48}$$

$$K = \frac{C_{S,2}}{C_{G,2}} = \frac{W_{S,2}}{W_{G,2}} \cdot \beta \tag{2.49}$$

where $W_{S,2} < W_{S,1}$; $W_{G,2} < W_{G,1}$; $C_{S,2} < C_{S1}$; $C_{G,2} < C_{G,1}$.

If we were to withdraw the same aliquot volume from the headspace of the vial, the peak area obtained (A_2) would be smaller than area A_1 obtained in the first analysis.

Although the foregoing derivation is generally true, the extent of the change depends very much on the analytical conditions: that is, on the partition coefficient and the phase ratio. The extent of the change can be expressed by a quotient Q, representing the ratio of the peak areas obtained in the two consecutive determinations and, due to the proportionality between peak area and concentration, this quotient will also be equal to the ratios of the respective headspace concentrations:

$$Q = A_2/A_1 = C_{G,2}/C_{G,1} \tag{2.50}$$

Expressing C_G from eq. 2.19, we can write

$$C_{G,1} = \frac{C_{o,E1}}{K + \beta} = \frac{W_{o,1}}{V_S \cdot (K + \beta)} \tag{2.51}$$

$$C_{G,2} = \frac{C_{o,2}}{K + \beta} = \frac{W_{o,2}}{V_S \cdot (K + \beta)} = \frac{W_{o,1} - W_A}{V_S \cdot (K + \beta)} \tag{2.52}$$

and

$$Q = \frac{A_2}{A_1} = \frac{C_{G,2}}{C_{G,1}} = \frac{W_{o,1} - W_A}{W_{o,1}} \tag{2.53}$$

where $W_{o,1}$ is equal to W_o, the original amount of the analyte present in the sample.

Later, in Section 5.5.4, we shall deal in detail with the role of quotient Q in consecutive analyses representing the basis of the multiple headspace extraction technique, the theoretical basis of which is discussed below. Let it suffice to say that the value of Q (which is always less than unity) depends on the value of K/β. If this value is close to or less than unity, the value of Q will be small; on the other hand, if K/β is large (above about 4), the value of Q approaches unity. This means that in this case, the change in the analytical results from two consecutive analyses from the same vial would be very small. This is the case, for example, in the determination of ethanol in water (or blood), where the partition coefficient is very high. On the other hand—for example, in the analysis of halogenated hydrocarbons in water—the change would be significant because the partition coefficient of these substances are low (cf. Table 2.2).

The practical conclusion of these considerations is that in general practice, one should never sample twice from the same vial because the magnitude of the inherent difference in the two determinations cannot be predicted. Therefore, if duplicate analyses are desired, one must place the *same* sample with the same volume in two *separate* vials and analyze them consecutively, under identical conditions.

2.6 Multiple Headspace Extraction (MHE)

2.6.1 Principles of MHE

We have discussed the situation of two consecutive aliquots taken from the headspace of the same vial: although the partition coefficient K remains constant, the peak area obtained for the second aliquot may be smaller than the area obtained in the first analysis. If we continue taking successive aliquots from the vial's headspace, the total amount of the analyte present will further decline, eventually becoming totally exhausted. Thus, the sum of the amounts of the analyte removed in the individual extractions will be equal to the total amount of analyte present in the original sample. This is what we call the multiple headspace extraction method. Its advantage is that by extracting the *whole* amount of the analyte, any effect of the sample matrix is eliminated and quantitative determination of the total amount present depends only on the relationship (i.e., on the response factor value), just as in any quantitative measurement by GC.

In practice we do not carry out the extractions in MHE ad infinitum: from a limited number of consecutive extractions the peak area corresponding to the total amount of analyte present is obtained by extrapolation, based on mathematical relationships.

In principle, multiple headspace extraction is *dynamic gas extraction carried out stepwise*. We may compare it to repeated liquid extractions in a separation funnel: in each step, part of the analyte present is removed until no analyte is left in the original sample.

The MHE technique and the mathematical model was originally described by McAuliffe [13] and Suzuki et al. [14], but a number of other scientists [15–19] also contributed to the development of its theory. Kolb and Ettre [20] compared the empirical and exact theories of MHE, demonstrating that these give the same results, thus refuting the objections of Venema [21].

2.6.2 Theoretical Background of MHE

As mentioned, in MHE, we carry out consecutive analyses from the same sample vial. After each sampling, part of the remaining headspace is removed so that the pressure in the vial returns to atmospheric (or close to it) and then the vial is reequilibrated. During these consecutive analyses we obtain peak area A_1, A_2, etc.: the sum of these peak areas (the *total area* in the following tables) is proportional to the total amount of the analyte originally present in the sample (W_o):

$$\sum_{i=1}^{i=\infty} A_i = A_1 + A_2 + \cdots + A_i \tag{2.54}$$

$$W_o \propto \sum_{i=1}^{i=\infty} A_i \tag{2.55}$$

The decline in the peak area values is exponential: Figure 2.10 shows the chromatograms from three consecutive measurements of an aqueous sample containing volatile halogenated hydrocarbons [22].

To establish the sum of the peak area, let us consider first the situation in a continuous gas extraction for which at first order mechanism is applied as shown in eq. 2.56. Here the concentration-versus-time (C vs. t) relationship can be described as:

$$\frac{dC}{dt} = qC \tag{2.56}$$

where q is a constant which describes how fast the extraction process proceeds. At a given time t the concentration C depends on the initial concentration C_o and on the exponent q:

$$C = C_o \cdot e^{-qt} \tag{2.57}$$

If now the process is carried out stepwise (as in MHE) and each time, the peak area corresponding to the analyte concentration in the extract (in the headspace) at that step is determined, we can replace t by the number of extraction steps i and C by the peak area obtained A_i:

$$A_i = A_1 \cdot e^{-q(i-1)} \tag{2.58}$$

In eq. 2.58 we use $(i-1)$ for t and A_1 for C_o, because the first extraction takes place at $t = 0$. We can thus write eq. 2.54 in the following way:

$$\sum_{i=1}^{i=\infty} A_i = A_1 + A_1 \cdot e^{-q} + A_1 \cdot e^{-2q} + \cdots + A_1 \cdot e^{-(i-1)q} \tag{2.59}$$

$$\sum_{i=1}^{i=\infty} A_i = A_1[1 + e^{-q} + e^{-2q} + \cdots + e^{-(i-1)q}] \tag{2.60}$$

This is a converging geometrical progression and enables us to derive the sum of the individual members as follows:

$$\sum_{i=1}^{i=\infty} A_i = \frac{A_1}{1 - e^{-q}} \tag{2.61}$$

In other words, we can calculate the sum of all peaks from two values: the peak area obtained in the first extraction A_1 and the exponent q.

The exponent q describes the exponential decline of the peak areas during the stepwise MHE procedure and can thus be derived by eq. 2.58 from the area ratio Q of the consecutive peaks:

$$Q = \frac{A_2}{A_1} = \frac{A_3}{A_2} = \frac{A_{(i+1)}}{A_i} = e^{-q} \tag{2.62}$$

To obtain q, we write eq. 2.58 in the following form:

$$\ln A_i = -q(i-1) + \ln A_1 \tag{2.63}$$

This is a linear equation of the $y = ax + b$ type, where $x = (i-1)$, $y = \ln A_i$; the slope (a) is $-q$ and the y-intercept is $\ln A_1$.

The conclusion of this derivation is that in MHE, a few consecutive measurements and the linear regression analysis of the data according to eq. 2.63 [$\ln A_1$ vs. $(i-1)$] gives the quotient Q:

$$-q = \ln Q \tag{2.64}$$

$$Q = e^{-q} \tag{2.65}$$

From its value and the peak area obtained in the first extraction (A_1) the sum

of all peak areas corresponding to the total amount of the analyte present in the sample (W_o) can be calculated:

$$\sum_{i=1}^{i=\infty} A_i = \frac{A_1}{1 - e^{-q}} = \frac{A_1}{1 - Q} \tag{2.66}$$

2.6.3 Simplified MHE Calculation

If the linear regression calculation shows a very good correlation, or, if on the other hand, very high accuracy is not required, the determination can be carried out from the first two consecutive measurements only ("two-point calculation"). As we have seen (eq. 2.62):

$$Q = e^{-q} = A_2/A_1$$

Substituting this expression into eq. 2.61 we obtain:

$$\sum_{i=1}^{i=\infty} A_i = \frac{A_1}{1 - (A_2/A_1)} = \frac{A_1^2}{A_1 - A_2} \tag{2.67}$$

How close the result of the two-point calculation is to the result obtained in multipoint linear regression depends very much on how close the ratio of the first two area values is to the ratio of the subsequent pairs: even a small random deviation may result in a notable difference in the results calculated in the two different ways.

The same problem exists with the standard addition technique, if carried out by a single addition (cf. Section 5.4.1). Let us not forget that one single ratio is used here to calculate the quotient Q, while in multipoint measurement, calculation of the slope (and from it, Q) is done by linear regression analysis which compensated for random variations. Therefore, the simplified, two-point calculation should only be used in routine analysis, after the closeness of this approach to the result of regression analysis has been checked from multipoint measurements.

References

1. Howard Purnell, *Gas Chromatography*, Wiley, New York and London, 1962; pp. 9–31.
2. C. D. Hodgman, R. C. Weast, and S. M. Selby, *Handbook of Chemistry and Physics*, 42nd ed., Chemical Rubber Publishing, Cleveland, 1960: (a) pp. 2326–2329; (b): pp. 2448–2455.
3. E. F. Herington, in D. H. Desty (editor), *Vapour Phase Chromatography (1956 London Symposium)*, Butterworths, London, 1957; pp. 5–14
4. L. S. Ettre, J. E. Purcell, J. Widomski, B. Kolb, and P. Pospisil, *J. Chromatogr. Sci.* **18**, 116–124 (1980).
5. B. Kolb, C. Welter, and C. Bichler, *Chromatographia*, **34**, 235–240 (1992).
6. L. S. Ettre, C. Welter, and B. Kolb, *Chromatographia*, **35**, 73–84 (1993).
7. L. S. Ettre and B. Kolb, *Chromatographia*, **32**, 5–12 (1991).
8. R. J. Steichen, *Anal. Chem.* **48**, 1398–1402 (1976).
9. H. Hachenberg and A. P. Schmidt, *Gas Chromatographic Headspace Analysis*, Heyden & Son, London, 1977: (a) pp. 13–15; (b) pp. 82–116.

10. B. V. Ioffe and A. G. Vitenberg, *Headspace Analysis and Related Methods in Gas Chromatography*, Wiley-Interscience, New York, 1984; pp. 23–24.
11. B. Kolb, *J. Chromatogr.* **112**, 287–295 (1975).
12. B. Kolb, in B. Kolb (editor), *Applied Headspace Gas Chromatography*, Heyden & Son, London, 1980; pp. 1–11.
13. C. McAuliffe, *Chem. Technol.* **1971**, 46–51.
14. M. Suzuki, S. Tsuge, and T. Takeuchi, *Anal. Chem.* **42**, 1705–1708 (1970).
15. J. Novák, *Quantitative Analysis by Gas Chromatography*, M. Dekker Inc., New York, 1975; pp. 107–156.
16. B. V. Ioffe and A. G. Vitenberg, *Chromatographia*, **11**, 282–286 (1978).
17. J. Drozd and J. Novák, *J. Chromatogr.* **285**, 478–483 (1984).
18. B. Kolb, *Chromatographia*, **15**, 587–594 (1982).
19. A. G. Vitenberg and T. L. Reznik, *J. Chromatogr.* **287**, 15–27 (1984)
20. B. Kolb and L. S. Ettre, Chromatographia, **32**, 505–513 (1991).
21. J. Venema, *High Resolution Chromatography and Chromatography Communications HRC/CC*, **9**, 637–649 (1986).
22. B. Kolb, M. Auer, and P. Pospisil, *Gewässerschutz, Wasser, Abwasser*, **57**, 101–125 (1982).

The Techniques of Headspace–Gas Chromatography

In this chapter we deal with the sample vials and the systems in which the analysis is carried out. Essentially, headspace–gas chromatography can also be carried out manually, using a gas-tight syringe. ASTM and the U.S. Pharmacopeia still list this possibility among the officially accepted methods. However, manual operation is very time-consuming and labor intensive. Most laboratories today use fully automated systems that can control the thermostat in which the samples are equilibrated. Such automated systems also provide much better control of the analytical conditions and thus assure their reproducibility. We pointed out in the discussion of the theoretical background of HS-GC that this is a prerequisite for accurate quantitative analysis.

We shall only summarize the principles of the automated systems, since their construction and operation are covered by company literature, and the details vary from model to model. However, we shall deal in more detail with questions associated with the disposable items selected by the analyst (vials, septa) and with the special techniques (backflushing and cryogenic sample trapping) used in HS-GC. These questions are surprisingly little known and are covered scarcely at all in the literature.

3.1 Sample Vials

3.1.1 Types

In manual operation, essentially any small vessel can be utilized. Early investigators actually used standard laboratory Erlenmeyer flasks closed with a serum cap, taking out an aliquot of the headspace gas with a syringe.

Analysts today use standardized vials made of borosilicate glass, with 5–22 mL nominal volumes, and these are available from a number of suppliers. It is emphasized that the thermostat of each commercial headspace instrument requires a certain type of vial, and its specifications must be considered when the vials for a given instrument are purchased.

The vials for HS-GC analysis are produced in large quantities, with a constant volume. The specified nominal volume (which is a rounded-up value) is usually very close to the actual volume. Since in many quantitative methods (see Chapter 5) the actual vial volume must be known (to establish the phase ratio), it is advisable to determine this value if the work at hand involves a new batch or a new supplier. This can be done by, for example, completely filling a number of vials with water, weighing them, and then using the mean value for future calculations. In one case, a batch of 10 vials having a nominal value of 22 mL showed a mean value of 22.331 mL with a relative standard deviation of $\pm 0.35\%$.* The volume of the vials was calculated from the density of water (0.9971 g/mL at 25 °C).

Figure 2-1 has already shown a typical version of a headspace vial. These vials are equipped with a septum and sealed with an aluminum cap (see Figure 3-1). Care is necessary for a good design of the upper rim at the top of the vial to provide for a tight seal. This is important with flat, disk-type septa, which must be pressed tightly onto the top of the vial. If the top is flat also, the glass surface may be uneven, containing scratches and grooves, which then may cause a leaking seal. This is particularly important with aluminum- and Teflon-lined septum disks, because the protecting layer must be thick enough to provide the necessary diffusion barrier and inertness. Since, however, a thick protecting layer is likely to be too hard to be pressed smoothly onto the glass top of a vial if this area is flat, some vials have an angular upper rim in the form of an obtuse edge at the top (*A2*), which is pressed strongly against the flat septum when crimping the aluminum cap [1].† Other vials have a bevelled top (*A3*) [2] for similar reasons. On the other hand, however, this type of pressure-tight closure requires provisions for a safe pressure release (cf. Section 3.2.2) in case of a dangerous buildup of high pressure.

3.1.2 Selection of the Vial Volume

As mentioned, the commercial instruments use vials of certain types. Their size is generally a compromise. For the determination of residual solvents in printed films by HS-GC, a large aliquot is required to get a representative pattern, and therefore relatively large vials are needed. On the other hand, in

---

* These results are very consistent with the vials used in our laboratory. Therefore in all calculations in this book, we used 22.3 mL as the volume of the vial, which is the mean of the determined values.

† In the Perkin-Elmer headspace samplers, these vials are used together with the safety closure (Section 3.2.2).

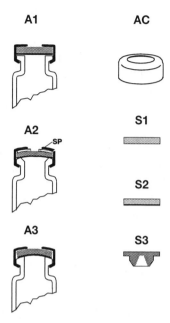

Figure 3-1. Vial systems and accessories used with them. *A1* = vial with flat top, *A2* = vial with an angular top (*SP* = star spring), used in the Perkin-Elmer safety closure system (see Section 3.2.2), *A3* = vial with a beveled top. *AC* = aluminum cap. Septa: *S1* = regular septum disk, *S2* = septum disk with Teflon or aluminum coating, *S3* = stopper-type septum.

the case of liquid samples one does not need a large vial. A smaller sample volume allows shorter equilibration times. The headspace sensitivity primarily depends on the concentration in the gas phase and not on the sample size or vial volume: it is the phase ratio β that determines the final sensitivity. Therefore, a 2.5 mL liquid sample in a 10 mL vial gives the same sensitivity as a 5 mL sample in a 20 mL vial (the phase ratio β in both cases is 3).

With packed columns, the headspace gas injected into the column is comparable to the volumes of gas samples used in general gas chromatography, in the order of 0.5–2 mL. In this case a larger vial is needed. In the case of open-tubular columns with split injection, the situation is similar to packed columns. With splitless sample introduction, on the other hand, the headspace gas entering the column is only in the range of 25–250 μL, and thus here a small vial, with a small sample volume, is sufficient.

This discussion demonstrates the many aspects of size selection for HS-GC vials. Vials with nominal volumes around 20 mL are considered to be a reasonable compromise. Some instruments can use vials with two different volumes, a smaller and a larger.

3.1.3 Vial Cleaning

In some methods precleaning of the vials (by washing with a detergent solution followed by rinsing with distilled water and drying in an oven) is recommended. However, according to our experience, the headspace vials as obtained are sufficiently clean and no pretreatment is necessary. In fact, such washing, particularly with a detergent, may add impurities to the vial!

The main source for blank peaks from an empty vial is not contamination on the vial's inner surface but bleeding of the septum (see Section 3.3) and/or impurities in the air filling the vial. This was demonstrated by Sadowski and Purcell [3], who detected the presence of halogenated hydrocarbon contaminations due to atmospheric pollution from a neighboring plant using chlorinated solvents. Such volatile contaminants can be eliminated by storing the empty vials in a clean room and/or by purging them with an inert, pure gas before the sample is added to the vial.

3.1.4 Wall Adsorption Effects

Can analyte loss occur as a result of adsorption on the inner surface of the glass vial? It is certainly possible, and if such loss were experienced, it could be observed by means of a nonlinear MHE plot, particularly at lower concentrations. On the other hand, the MHE plots shown in this book generally have a good linearity (except if some other problems are involved), indicating the absence of any analyte loss due to wall adsorption. In our opinion, this finding may be explained by the presence of a permanently adsorbed water layer on the inside glass surface due to hydrogen bonding to the surface silanol groups. The water originates from the humidity of the atmosphere present in the vial (cf. Table 2-1), and from the natural humidity of most samples. In this way, the glass surface does not act as an adsorbent toward the analyte molecules: in fact, it now acts as a partition system that is stable up to about 120–150 °C when the water layer begins to be stripped off. Only if the vial's temperature exceeds 150 °C will the wall adsorption effects become significant. The following example demonstrates this.

Example 3.1

The aim was to analyze ethylene glycol (EG), which is a very polar and fairly high-boiling compound (boiling point: 197.6 °C). An attempt to analyze EG by HS-GC from an aqueous solution failed, as a result of the compound's very high solubility in water and the corresponding low concentration in the headspace. Therefore, the use of the total vaporization technique (TVT: see Section 4.6.1) was considered. Specifically, 5 μL of the aqueous sample was introduced into the vial, which was heated to 180 °C, at which temperature the EG completely evaporated. A reasonable EG peak was obtained; however, when the MHE procedure was used for quantitative analysis, the plot was highly nonlinear, showing a dramatic loss at lower concentrations (see Figure 3-2, plot A). First we tried to silanize the glass vial (a procedure generally considered to deactivate a glass surface), but this did not yield

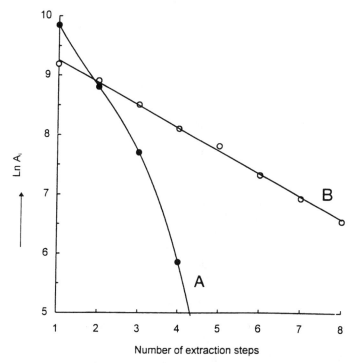

Figure 3-2. MHE plots of ethylene glycol, thermostatted at 180 °C. $A = 5\,\mu$L aqueous ethylene glycol (10%) solution; $B =$ same as A, but with one drop (about 30 mg) glycerol added with the sample to the vial ($r = 0.9988$). For conditions, see Figure 3-3.

any improvement. However, the addition of one drop of glycerol helped, resulting in a linear MHE plot (see Figure 3-2, plot B). Apparently the vapor of glycerol (BP: 290 °C) already covered and deactivated the glass surface at 180 °C. This technique permitted the quantitative analysis of an aqueous extract of a soil sample with 50 ppm EG concentration (see Figure 3-3).

3.2 Caps

The headspace vial is closed by a septum and a cap. Screw-cap closures have been used mainly in manual systems, but in today's automated instruments the cap is crimped tightly over the septum and the top of the vial. Both manual and automated or semiautomated crimping devices are available. The crimper must be adjusted carefully, particularly if flat septum disks are used, while rubber stoppers are less critical. A poorly adjusted crimper is often the source of a leaking seal. The tightness of the seal can be checked by trying to rotate the cap by hand: this should not be possible or should require considerable

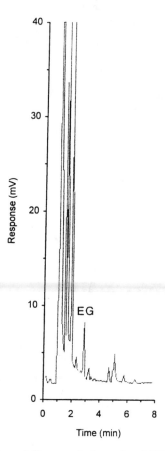

Figure 3-3. Determination of 50 ppm ethylene glycol (EG) in an aqueous soil extract. *HS conditions*: Sample: 10 μL aqueous solution to which one drop (30 mg) of glycerol was added. Equilibration at 180 °C for 30 minutes. Sample transfer time: 4.8 seconds. Calibration by external standard. *GC conditions*: Column: 25 m × 0.25 mm I.D. fused-silica, open-tubular, coated with methyl silicone stationary phase; film thickness: 1 μm. Column temperature: isothermal at 60 °C. Split injection. Flame-ionization detector.

strength. In another test, a few microliters of a highly volatile compound (e.g., *n*-pentane) is added to the vial which, after sealing, is dipped in hot water: if *n*-pentane leaks out and escapes, gas bubbles will be observed.

With an automated instrument, a leaking vial can also be recognized if one places the same homogeneous vapor sample (by TVT) in a few vials, using different pressurization times but the same sample transfer time. In the case of a hermetic seal, the peak area should remain constant: a decreasing peak area with increasing pressurization time indicates leaks.

The caps are generally made of aluminum, and various types are available. In some instruments caps made of a special polished steel alloy are picked up by a magnet for placement into the thermostat [2].

3.2.1 Pressure on Caps

Modern automated instruments allow vial temperatures well above 100 °C. In such cases it is almost mandatory to prevent excessive pressure in the vials, which could cause explosive breakage of the glass vial. Such catastrophic failures not only may hurt the operator but also may damage the instrument. The actual pressure at which this can occur depends on the thickness of the glass wall: generally, it is above 10^3 kPa.

The unsafe high pressure formed in the vial during thermostatting may be due to the use of a solvent that is too low boiling, but usually it is simply due to operator error. For example, if an aqueous sample is thermostatted to 80 °C, the internal water vapor pressure is only 47 kPa. If, however, the operator inadvertently enters an erroneous value of 180 °C, the internal pressure will be as high as 10^3 kPa, with a high risk of vial explosion.

3.2.2 Safety Closures

Today, some vial caps are available with special crimp closures, that enhance safety by incorporating bridges and scorelines in the aluminum seal [2]. Figure 3-4 shows the construction of the closures of the vials supplied by the

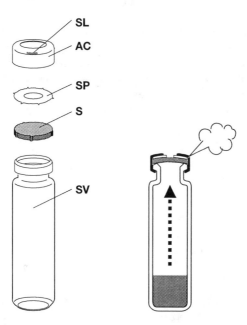

Figure 3-4. Safety closure for headspace vials used in Perkin-Elmer instruments [1], opening an artificial leak at pressures > 500 kPa. *SV* = sample vial with angular top (22.3 mL and 9.6 mL internal volume), *S* = septum disk, *SP* = star spring, *AC* = aluminum cap with *SL* slot.

Perkin-Elmer Corporation [1]: here the internal pressure in the vial presses the septum disk against a star spring and the aluminum cap, which has a small, curved slot. At pressures above 500 kPa, this slot is deformed and an artificial leak opens, venting the internal pressure to atmosphere. It should be mentioned that this safety closure requires flat septum disks and will not work with rubber stoppers, which therefore are not permissible for this safety arrangement.

3.3 Septa

3.3.1 Types

Today a wide variety of septa are available for HS-GC and Table 3-1 lists various types illustrated in Figure 3-1 (*S1–S3*). Selection of a certain type usually represents a compromise between price and the specific requirement of an application, particularly temperature stability.

The low-cost butyl rubber seals are of limited use, since they rapidly absorb nonpolar compounds and have a relatively low upper temperature limit. On the other hand, polar compounds, particularly alcohols, showed negligible loss with these seals. This is an important point in forensic laboratories carrying out a tremendous number of analyses, where cost is an important consideration [4].

Most commonly Teflon-lined butyl rubber septa are used, despite their limited temperature maximum (about 100 °C). The best, and most inert protecting shield is provided by the aluminum-lined septa; their only shortcoming is that they often also include a polyethylene layer, which limits the upper temperature to about 120 °C. However, aluminum-lined silicone rubber septa permitting higher temperature operation are also available.

Table 3-1   Characteristics of the septa available for HS-GC

Label in Figure 3-1	Type	Upper temperature limit (°C)	Inertness	Price
S1	Butyl rubber disk	100	poor	low
S2	Teflon-lined butyl rubber disk	100	good	medium
S2	Aluminum-lined silicone rubber disk	120* (200)	good	medium
S2	Teflon-lined silicone rubber disk	210	good	medium
S3	Viton rubber[†] stopper	200	good	high
S3	Butyl rubber stopper	80	poor	low

* This limit concerns aluminum-lined septa, which also include a thin polyethylene layer.
[†] Viton is a vinylidene fluoride/hexafluoropropylene copolymer, a product of E. I. DuPont de Nemours & Co., Inc.

With respect to the temperature limit of the septa, it is important to emphasize that it applies not only to the temperature of the vial, but also to the temperature of the instrument's needle used for pressurization and sample transfer, which is heated to prevent condensation. If, for example, the vial is at 80 °C, but the needle is heated to 150 °C, the hot needle may decompose the septum material, resulting in spurious peaks in the chromatogram and a leak around the needle. This may also occur with laminated septa, where one of the layers may have a lower temperature limit than the bulk material or vice versa.

3.3.2 Septum Blank

All septa have some residual lower-boiling compounds present, either from the manufacturing process or from adsorption during storage or use. Whether these compounds will be detected depends mainly on detector sensitivity. Figure 3-5 illustrates the analysis of trace concentrations of volatile halogenated hydrocarbons in water, with an electron-capture detector, together with a blank run (B). Peak 5 is from 0.1 ppb (μg/L) 1,1,1-trichloroethane: this compound is frequently found as a ubiquitous blank in water and air. The corresponding peak in the blank represents about 8 ppt (ng/L). However, the source of these blank peaks is not necessarily the septum. Pure water had to be analyzed for the blank, and it may be more difficult to get such "pure" water than to get a clean septum. In practice it is almost impossible to localize the source of such small peaks, whether the corresponding impurity comes from the septum, the "pure" matrix used in the blank measurement, or from the laboratory atmosphere. For example, the blank peak corresponding to dichlorobromomethane (peak 9) representing a concentration of about 15 ppt, definitely originated from the air present in the laboratory on that particular day. For identification of all the peaks in Figure 3-5, see Table 3-2.

Table 3-2 Identification of the peaks in Figure 3-5

Peak no.	Compound	Concentration (μg/L) (ppb)
1	1,1-Dichloroethylene	5.9
2	Dichloromethane	9.0
3	1,1-Dichloroethane	8.0
4	Chloroform	1.4
5	1,1,1-Trichloroethane	0.1
6	1,2-Dichloroethane	11.9
7	Carbon tetrachloride	0.07
8	Trichloroethylene	0.7
9	Dichlorobromomethane	0.2
10	Dibromochloromethane	0.2
11	Tetrachloroethylene	0.16
12	Bromoform	1.4

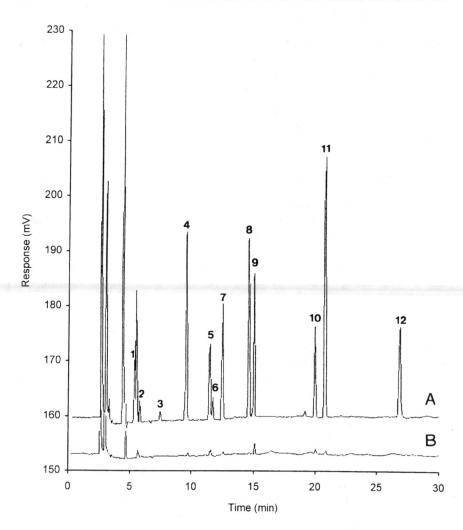

Figure 3-5. Analysis of volatile halogenated hydrocarbons in water by HS-GC, using an electron-capture detector. *A* = chromatogram of the sample, *B* = blank run with 5 mL of pure water in the vial. *HS conditions*: Sample: 5 mL, thermostatted at 80 °C for 30 minutes with shaker. Sample transfer time: 2.4 seconds. *GC conditions*: Column: 50 m × 0.32 mm I.D. fused-silica, open-tubular, coated with bonded phenyl (5%) methyl silicone stationary phase; film thickness: 2 μm. Column temperature: isothermal at 50 °C for 10 minutes, then programmed at 8 °C/min to 100 °C. Splitless injection. Carrier gas: Nitrogen, 140 kPa. Electron-capture detector; with 50 mL/min nitrogen as makeup gas. For peak identification see Table 3-2.

If trace analysis is to be carried out, all the headspace accessories (vials, septa) should be stored in a clean environment, away from the laboratory, and sample preparation should also be carried out in a "clean" room.

The blank run shown in Figure 3-5 shows only minimum interference; still, such small peaks may limit the use of any further enrichment techniques — for example, cryofocusing — with such a highly sensitive detector. Other, less sensitive detectors, such as the flame-ionization detector (FID), would show no blank peaks and would allow further enrichment by cryofocusing (see Section 3.8). Figure 3-6 illustrates the blank of four different septa at high FID sensitivity, obtained by a sample transfer time of 2 minutes and cryogenic sample trapping. As seen, the silicone rubber septum with Teflon liner (A) had practically no blank, and the aluminum-coated silicone rubber septum (B) only had two additional peaks at the high temperature range. However, the uncoated butyl rubber septum (D) is unacceptable for such an application.

But what can be done if the available septa become contaminated? The best advice is to discard them and obtain a new batch. If, however, the septa must be cleaned, the following possibilities may be tried:

- heating the contaminated septa in a glass container, with continuous flushing with an inert gas, to the recommended maximum temperature for several days
- boiling them in water while purging with an inert gas
- spreading the septa in a drawer in a clean environment for a long time (maybe several weeks), and waiting for self-cleaning by slow release of the contamination.

3.3.3 Should a Septum Be Pierced Twice?

Is it safe to puncture a septum more than once? This important question is often asked, for dual piercing may be done when an internal standard, a small volume of a liquid sample, or a gas sample is added with a syringe through the septum, into an already closed vial, and it may also be used in some systems that carry out MHE by removing the sample needle each time after venting, before reequilibration, and then reintroducing it for the next pressurization and sample transfer. In this respect it is important to realize that with most automated headspace samplers the (outside) diameter of the sampling needle is larger than the diameter of the needle of the usual microliter syringe.

There is no unambiguous answer to this question, which depends on many practical parameters and the properties of the sample. In general, however, we must consider two questions: pressure tightness and diffusion tightness. The first refers to loss of pressure in the vial because of a leak. This may happen during the pressurization step used in both the balanced pressure and pressure/loop systems, if the septum has already been pierced once. When the needle breaches the septum for the second puncture, the septum may become mechanically deformed and the first puncture, although still tight up to this

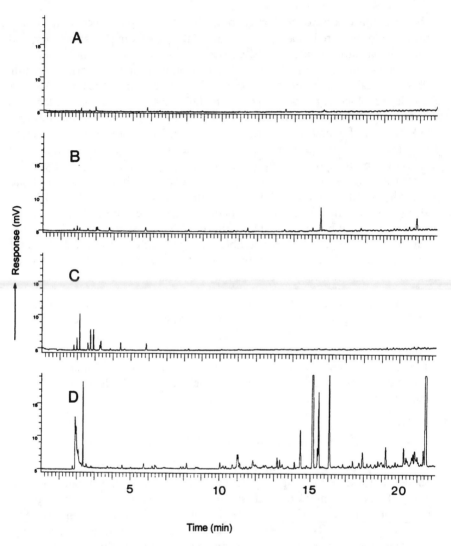

Figure 3-6. Blank runs of four different septa, by cryofocusing HS-GC, with a flame-ionization detector. A = Teflon-coated silicone rubber, B = aluminum-coated silicone rubber, C = Teflon-coated butyl rubber, and D = uncoated butyl rubber. *HS conditions*: Sample: empty vials flushed with helium before closing, thermostatted at 100 °C for 30 minutes. Needle temperature: 100 °C, sample transfer time: 2 minutes with cryofocusing (cf. Figure 3-22). *GC conditions*: Column: 50 m × 0.32 mm I.D. fused-silica, open-tubular, coated with bonded phenyl (5%) methyl silicone stationary phase; film thickness: 1 μm. Column temperature: isothermal at 50 °C for 3 minutes then programmed at 8 °C/min to 200 °C. Carrier gas: Helium, 180 kPa.

instant, may be deformed also, possibly creating a leak. Loss of pressure tightness is particularly likely to be observed if the vial contains a homogeneous vapor (using the TVT: see Section 4.6.1).

The second criterion, diffusion tightness, is especially likely to be a problem with Teflon- or aluminum-lined septa. When these are punctured, the protecting layer is destroyed and through the resulting small hole, the unprotected septum material is now accessible to absorb volatiles from the headspace vial. Such loss depends on the time between the first piercing (injection of a sample through the septum into the already closed vial) and the actual analysis, on the sample properties, and on the permeability of the septum material. The problem becomes crucial with homogeneous vapor mixtures that are stored in the vial for days after the septum has been punctured.

Table 3-3 gives some interesting data on such possible loss due to diffusion. Here, a standard solution of acetone (1%) and dichloromethane (0.2%) in dimethylformamide was prepared and 3 μL of this solution was added to a number of vials by two techniques: by piercing through the septum of the already closed vial with the syringe needle, and by the open-vial technique (described shortly). The vials were stored for 3 days, then analyzed by HS-GC and the peak areas compared: the results from the nonpunctured vial were always taken as 100. As seen in Table 3-3, in the vials closed with Teflon-lined silicone rubber septa, 69% of acetone and 89% of dichloromethane were lost, while in the case of the Teflon-lined butyl rubber septa, the losses were only 2.6%. These results indicate that if a sample is introduced into an already closed vial by syringe injection, it should be analyzed promptly. The butyl rubber apparently is more dense and less permeable than the silicone rubber, and these results also indicate that from this point of view, the Teflon-lined butyl rubber septum is preferable to the Teflon-lined silicone rubber septum.

Table 3-3 Comparison of septum tightness for vapor samples*

| | Relative peak area | | | |
| | Teflon/silicone rubber septum | | Teflon/butyl rubber septum | |
Compound	Unpunctured septum	Punctured septum	Unpunctured septum	Punctured septum
Acetone	100 (± 1.3%)	31 (± 4.9%)	100 (± 0.8%)	97.4 (± 3.5%)
Dichloromethane	100 (± 1.3%)	11 (± 8.2%)	100 (± 1.0%)	97.4 (± 3.5%)

*3 μL of a DMF solution was introduced into a series of vials, either by puncturing the septum of the closed vial or by the open-vial technique, and the vials were stored for 3 days in a refrigerator before analysis. *HS conditions*: equilibration at 80 °C for 30 minutes; needle temperature: 90 °C, pressurization at 110 kPa for 3 minutes. The peak area of the unpunctured vials is taken as 100. Relative standard deviation is for *n* = 5 in each case.

As already stated, these problems occur primarily with vials in which the sample was prepared by the total vaporization technique by injecting a few microliters of the sample with a syringe into the already closed vial — in other words, when the vial is filled with the vapor of the sample. If, however, the vial contains a liquid or solid sample in which the volatile analyte is dissolved according to its partition coefficient, then the loss by diffusion through the pierced septum may be compensated more or less depending on the value of the partition coefficient, by further evaporation from the condensed phase into the headspace. Therefore, it might be possible to add in this way an internal standard to a solution in the vial containing an analyte with a high partition coefficient (e.g., aqueous solution of ethanol), although even there long-term storage of vials should be avoided if the septum has already been pierced.

The following test (similar to the test illustrated in Table 3-3) demonstrates the situation when the vial contains a solution. Again 3 μL of the same solution of acetone and dichloromethane were added, but now the vials were all closed with Teflon-lined silicone septa and already contained 1 mL of water. Two techniques were used for sample addition: piercing the septum of an already closed vial, and also the open-vial technique. These vials were analyzed after standing for 3 days at room temperature, and the results are presented in Table 3-4, again taking the area values from the nonpunctured septa as 100%. We see that there is practically no difference for acetone due its good water solubility and correspondingly high partition coefficient. The aqueous sample apparently is a sufficient concentrated reservoir to compensate for any sample loss in the gas phase. On the other hand, dichloromethane showed a significant loss of 63% due to its low partition coefficient. It, therefore, behaves like the pure vapor sample as used in the foregoing test (see Table 3-3).

This test, which points out conditions under which punctured septa may be used with no problems, shows also that the open-vial technique works equally

Table 3-4   Tightness of Teflon/silicone rubber
septum for liquid samples*

Compound	Relative peak area	
	Unpunctured septum	Punctured septum
Acetone	100 (±1.2%)	101 (±1.3%)
Dichloromethane	100 (±0.96%)	37 (±2.1%)

*3 μL of a DMF solution was introduced into a series of vias, each containing 1 mL of water and closed by a Teflon-lined silicone septum, either by puncturing the septum of the closed vial or by the open-vial technique. The vials were stored for 3 days at room temperature. Instrumental conditions are the same as given in Table 3-3. Relative standard deviation is for $n = 5$ in each case.

Figure 3-7. Introduction of a liquid sample into a headspace vial by the open-vial technique.

well in the hand of a skilled operator, in spite of the potential risk of sample loss when the sample is injected into the open vial.

Closed-Vial Versus Open-Vial Sample Introduction Technique. The preceding discussion concerned the introduction of a sample or an internal standard with help of a microliter syringe, by piercing the septum of an already closed vial. Again, this technique may be used when an internal standard is being added to a solution but should be avoided in the total vaporization technique.

A better way to introduce the small volumes of such samples into vials is the open-vial technique (see Figure 3-7). The aluminum cap, including the septum, is already loosely placed at the top of the vial, but not yet crimped: a small slot remains open, through which the needle of a microliter syringe is inserted so that its end touches the inside wall of the vial. The sample is injected, the needle is immediately withdrawn, and the closure is pressed onto the vial as fast as possible and crimped as usual.

3.4 Thermostatting

Prior to analysis, the vials are thermostatted until equilibrium is reached between the two phases. Present-day automated instruments have thermostats that accommodate a number of vials and process them sequentially. Today the system consists of a heated metal block or an air thermostat; older systems also used an oil bath. Usually the system also incorporates a carousel, a rack, or a chain to store a large number of vials ready for processing, which are then automatically placed into the thermostat at a preset time. A special feature provided by some systems is a shaker or a magnetic stirrer. As discussed later (Section 4.1.2.1), the use of such a device can significantly reduce the time needed for equilibration.

Special care is needed in the design of the thermostat to provide a high constancy of temperature; also, special consideration is recommended concerning the actual time the sample vial spends in the thermostat at the elevated temperature.

3.4.1 Influence of Temperature

It is very important that the thermostat has a highly precise temperature control. However, for quantitative analysis, the constancy and particularly the long-term stability are more important than the absolute accuracy of the temperature, since any deviation from the true temperature equally affects all compounds in the sample and the calibration standard. The situation is, naturally, different in the determination of thermodynamic functions (see Chapter 9), where the true temperature must be known. Present-day thermostatting systems specify temperature control as precise as $\pm 0.1\,^\circ\text{C}$.

It was demonstrated in Chapter 2 that temperature influences volatility, and thus the headspace sensitivity of various compounds, in different ways. This is due to two relationships. The first shows the influence of the partition coefficient K on the analyte's concentration in the headspace C_G, that is, on the peak area A obtained when analyzing an aliquot of the headspace:

$$A \propto C_G = \frac{C_o}{K + \beta} \tag{2.19}$$

where C_o = analyte concentration in the original sample and β = phase ratio, while the second describes the relationship between temperature, in kelvins, and the partition coefficient:

$$\log K = \frac{B'}{T} - C' \tag{2.34}$$

where B' and C' are substance-specific constants.

Returning to Figure 2-4, we see immediately that polar compounds such as ethanol, with good water solubility and a high value of K, are much more affected by small temperature variations than nonpolar compounds with small values of K, if determined in an aqueous solution. We can investigate now how a small temperature variation of $0.1\,^\circ\text{C}$ as specified earlier will influence the analytical result. As given in Table 2-2, the values of the two constants for ethanol in aqueous solution are:

$B' = 2205.3$

$C' = 3.910$

Table 3-5 Influence of the sample temperature on the partition coefficient K of ethanol and the concentration in the headspace C_G, assuming a sample volume of 5 mL and a vial volume of 22.3 mL ($\beta = 3.46$)*

Temperature (°C)	K	$K + \beta$	$C_{G,i}/C_{G(ref)}$	Difference (%)
69.9	329.54	333.00	0.9957	−0.43
70.0	328.12	331.58	1.0000	0.00
70.1	343.26	330.17	1.0043	+0.43

* The partition coefficient was calculated according to eq. 2.34 using the values specified in the text. The ratio of the gas phase concentrations $C_{G,i}/C_{G(ref)}$ was calculated according to eq. 3.2.

From these data the values of K at 69.9, 70.0, and 70.1 °C can be calculated: they are listed in Table 3-5. To evaluate how much these differences influence the actual analytical results, we consider the standard blood alcohol analysis (see Section 5.2), where 0.5 mL of blood is diluted to 2.5 mL which is then analyzed by HS-GC. With a vial volume of 22.3 mL, this gives a phase ratio of $\beta = 19.8/2.5 = 7.92$. With a constant initial concentration and a constant phase ratio, we can write for two samples differing only in the value of K (cf. eqs. 2.19 and 2.36):

$$\frac{A_1}{A_2} = \frac{C_{G,1}}{C_{G,2}} = \frac{K_2 + \beta}{K_1 + \beta} = \frac{\alpha_1}{\alpha_2} \tag{3.1}$$

where C_G is the concentration of the analyte in the headspace, A is the corresponding peak area, and

$$\alpha = \frac{1}{K + \beta} \tag{2.36}$$

If we consider sample 2 as the reference (in this case the one analyzed at 70.0 °C), we can write eq. 3.1 in the following way:

$$\frac{A_i}{A_{ref}} = \frac{C_{G,i}}{C_{G(ref)}} = \frac{K_{ref} + \beta}{K_i + \beta} = \frac{\alpha_i}{\alpha_{ref}} \tag{3.2}$$

Table 3-5 lists the data calculated in this way for the three cases. As seen, a ± 0.1 °C temperature difference resulted only in a $\pm 0.43\%$ difference in the headspace concentration (i.e., in the resulting peak area).

For analytes with very low partition coefficient values, the influence of small temperature variations will be even less. With the data of Table 2-2 we can make the same calculation for an aqueous solution of tetrachloroethylene. The resulting difference in the gas phase concentration for a 0.1 °C temperature change will be insignificant ($<0.01\%$).

While any specified accuracy of the thermostat is certainly true for the temperature as measured at the position of the sensor (thermocouple), it is not necessarily true for the temperature of the sample in the vial. As mentioned earlier, the true value of the temperature is not as important for analytical applications, while it should be known as accurately as possible for thermo-dynamic applications. This is, however, not easy, because in a well-insulated thermostat the only easy way to access the sample in the vial for direct temperature measurement, entails inserting a thermocouple, with its inherent, temperature-disturbing thermal conductivity.

For a practical need, the exact temperature in the vial may be determined with an accuracy of at least $\pm 1\,°C^*$ by using the vial as a melting point device. This is surprisingly simple if some crystals of a pure standard compound with a well defined melting point are placed in the vial. For example, crystals of naphthalene with a melting point of 80.55 °C were still solid at a nominal instrument temperature of 81 °C, partly molten at 82 °C, and completely molten at 83 °C. Devices such as colored strips for certain temperatures are available for a wide temperature range and may also be used to calibrate the accurate inner temperature of the vial and thus of the sample.

3.4.2 Working Modes

It has been noted that the thermostat of a typical modern instrument is capable of holding a number of sample vials, which then are analyzed sequentially. In older designs, these vials were loaded simultaneously into the thermostat, and therefore later samples were thermostatted longer than necessary. It is true that in general, the analytical results will not change if the actual thermostatting time is longer than the time needed for equilibration; still, since some samples may be sensitive to prolonged heating, overly long thermostatting should be avoided.

The other possibility is to analyze one vial and immediately place another one into the thermostat and thermostat it for its equilibrium time. However, usually the equilibration times are fairly long, generally longer than the analysis time. To postpone starting the equilibration of the second sample until completion of the analysis of the first would result in an unnecessary loss of time and sample throughput.

This problem is solved by the so-called *overlapping constant mode* of thermostatting (see Figure 3-8). The stored vials are loaded automatically (e.g., from a carousel into the thermostat) at a preset time, so that they are all heated for the same period. This working mode is particularly important for two reasons: first, it minimizes the main drawback of static HS-GC, which is the sometimes long equilibration time. In a series of samples, only the first has to wait for the whole equilibration time, while the next vials are sequentially transferred into the thermostat with a time lag that equals the chromato-graphic cycle time. Several vials may thus be present in the thermostat simultaneously depending on the ratio of thermostatting time (TT) to cycle

---

* Most automatic headspace samplers allow the digital entry of the temperature at 1 °C steps.

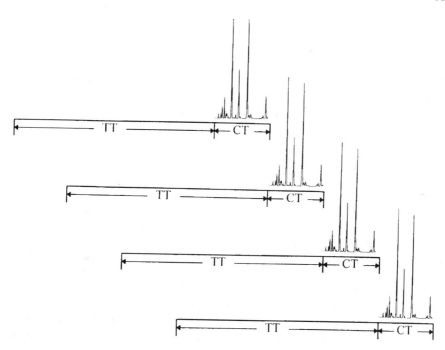

Figure 3-8. The overlapping constant working mode for thermostatting simultaneously several headspace vials: TT = thermostatting time, CT = chromatographic cycle time.

time (CT). When the analysis of the first vial is finished, analysis of the second vial starts immediately. Long equilibration times, therefore, do not affect the sample throughput in an automatic run of a series of samples. This working mode provides also for a constant equilibration time for all samples. This is important for samples that are not sufficiently stable over a longer time (diketones in beer, ethylene oxide in sterilized clinical material), where the conditions should be reproduced as closely as possible. In the case of different samples, one may set the longest thermostatting time: as already noted it is important only that the thermostatting time be at least as long as the minimum time to achieve equilibrium.

There is a second mode in selecting the thermostatting time. In this, the so-called *progressive mode*, the vials are thermostatted for an increasing time, representing a multiple of the first thermostatting time. Thus, for example, vial 2 is thermostatted for twice the time used for the first vial, vial 3 for three times that time, and so on.

The progressive mode is used in the determination of the necessary equilibration time, where a number of vials containing the same sample are to be thermostatted for increasing periods (see Section 4.1). The progressive mode is also useful in kinetic studies, such as investigating the progress of a reaction or the rate of release (cf. Section 9.6.2).

3.5 The Fundamental Principles of Headspace Sampling Systems

Headspace analysis can be carried out manually or using automated instruments. In essence, the existing systems correspond to three basic types: gas-tight syringes, the time-controlled, balanced pressure system, and the so-called pressure/loop systems.

3.5.1 Systems Using Gas Syringes

In the first sampling system we consider, an aliquot of the headspace is taken with a gas-tight syringe and then injected into the gas chromatograph, in the usual way. Naturally, manual injection can be used, and the ASTM standard practice describing the general procedure for analyzing volatile sample components by HS-GC [5] describes this possibility. However, manual syringe systems have two basic shortcomings. First, the pressure in the vial, hence the actual volume of the gas sample, is not controlled and actually changes during sample transfer because of expansion through the needle to atmospheric pressure. This problem can be partially eliminated by using gas syringes with pressure-lock (Luer-Lock) valves. The second problem is the uncontrolled temperature of the syringe, which can cause sample condensation inside the unit. The quoted ASTM standard practice recommends storing syringes in an oven at 90 °C between samplings, but of course there is no assurance of a constant temperature, and the problem of manually handling such a hot syringe remains to be addressed.

We should also mention here an interesting combination of a syringe with a gas sampling valve, used in a Russian laboratory for the analysis of wastewater from the manufacture of poly(vinyl chloride) [6]. The aim of this system was to eliminate the problems associated with an unheated gas syringe. The syringe is used only as a pump: the headspace aliquot is withdrawn from the sample vessel through the sample loop of a heated six-port valve; and the content of this heated sample loop is then injected into the gas chromatograph. Figure 3-9 illustrates the principles of this system.

Today, *automated systems* based on syringe injection are commercially available. Figure 3-10 illustrates the construction of an automated HS-GC system with syringe injection [7], including also the carousel in which the vials are stored prior to thermostatting and the thermostat, consisting of a heated metal block.* The vials are transferred automatically from the carousel to the thermostat at the appropriate (preset) time. At the end of the thermostatting time an aliquot of the headspace is taken by the (heated) gas syringe; then the syringe is withdrawn from the vial and moved above the injection port of the

---

* The principles of the carousel, a rack or a chain, and the thermostat are similar for automated systems of most types (except the way the vials are actually transferred from one to the other) and therefore, are shown only here.

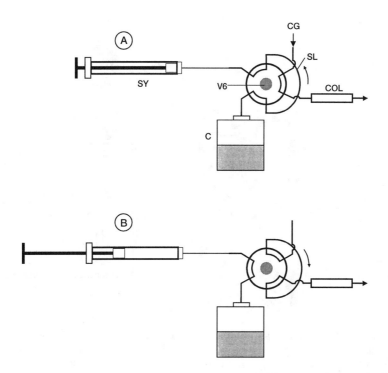

Figure 3-9. Combination of the use of a syringe with a heated six-port valve: After Litvinov et al. [6]. *A* = filling the sample loop, *B* = injecting the headspace aliquot into the column. *CG* = carrier gas, *V6* = heated six-port valve, *SL* = sample loop, *SY* = syringe, *C* = sample container, *COL* = column.

gas chromatograph, and the sample is injected. The principles of the system are essentially the same as those of the autosamplers for liquid sample injection which today are available for almost every gas chromatograph. In fact, Penton [8] described a modification of a regular autosampler to be used as a headspace sampler, by replacing the standard 10 μL (liquid) syringe with a 100 μL gas-tight syringe.

The syringes of the automated systems are usually heated; however, they do not have the possibility of locking the pressure within them. Thus the problems just mentioned still exist.

3.5.2 Balanced-Pressure Sampling Systems

In balanced-pressure systems an aliquot of the headspace of the vial is not withdrawn by suction as in the case of the syringe. Instead, after equilibrium

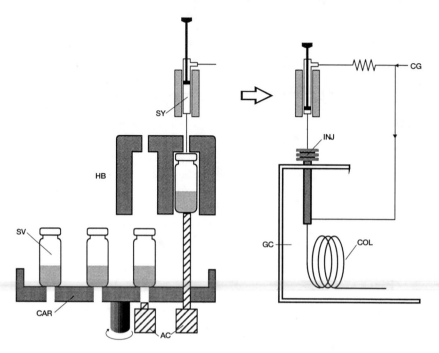

Figure 3-10. Principles of an automated HS-GC system with syringe injection [7]: SV = sample vial, CAR = carousel, HB = heated block, AC = air cylinder, SY = gas syringe, CG = carrier gas, GC = gas chromatograph, INJ = injection port of the gas chromatograph, COL = column.

has been reached, the vial is pressurized by the carrier gas to a pressure equal to the carrier gas inlet pressure of the column. Next, the carrier gas supply is interrupted by closing a valve in the carrier gas supply line and the pressurized gas in the vial expands onto the column, resulting in a flow of the mixed headspace gas from the vial to the column. Since both the pressure that builds up in the vial and the time of transfer can be set, the transferred volume of the headspace gas can be accurately controlled. In this system, representing the basic configuration for the "balanced-pressure" sampling technique, no separate gas is used for pressurization: the vial is pressurized to the column inlet pressure.

An automated system of this kind was first introduced in 1967 by the Perkin-Elmer Corporation [9]. In this system no rotating valve is used in the sample path, as shown by the schematic of its newer versions* [10] (Figure 3-11). A heated needle made of either stainless steel or an inert material (e.g., platinum), which has a hollow part permitting flow in either direction, moves

---

* Used in the Perkin-Elmer HS-100 and HS-40 automatic headspace samplers.

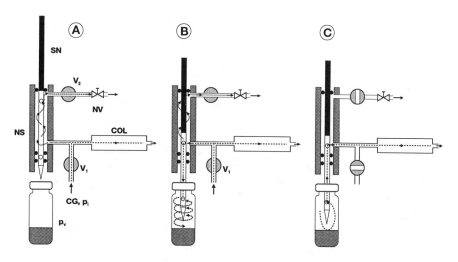

Figure 3-11. Schematic of the automated balanced pressure system [10]: (A) equilibration (standby), (B) pressurization, (C) sample transfer. CG = carrier gas, V = on/off solenoid valves, SN = movable sampling needle, NS = needle shaft, NV = needle valve, COL = column, p_i = column inlet pressure, p_v = original headspace pressure in the vial.

in a heated shaft that is continuously swept by a small purge gas flow to avoid any contamination. In the standby position (Step A, Figure 3-11A) the needle is sealed against atmosphere by O-rings. After the equilibration (Step A) has been completed, the needle penetrates the septum of the vial (Step B, Figure 3-11B), and part of the carrier gas flows into the vial to build up its pressure to the carrier gas inlet pressure. After a few minutes ("pressurization time") the carrier gas is temporarily disconnected by closing valve V_1 (Step C, Figure 3-11C). Since the sample vial is open to the column through the needle, we now have a flow of the headspace gas, transferring an aliquot of the gas into the injector through a heated inert transfer line. The volume of the aliquot is set by controlling the time of transfer. When this is accomplished, the system returns to the standby position (Step A, Figure 3-11A).

The term "balanced-pressure system" applies to the setup shown in Figure 3-11, where the pressure in the vial equals the head pressure of the column. It requires that the transfer line from the headspace sampler be connected directly to the column. For practical reasons, however, it is more convenient to connect an automated headspace sampler onto an existing injector of a gas chromatograph that has its own carrier gas supply providing the column head pressure p_i. To achieve the flow of the headspace gas into the injector and finally into the column, the headspace vial must be pressurized with an independent gas supply at a higher pressure p_p than the pressure in the injector ("*increased pressure sampling system*"). Such a modification was first described by Pauschmann [11] and Göke [12]. In the applicable figure captions, these pressure values are given as the ratio of pressurization gas to column inlet pressure, both in the pressure units as displayed by the instrument (gauge pressure), not as the absolute pressure values (e.g., Helium, carrier gas: 140:120 kPa, split sampling).

The system as shown in Figure 3-11 in its latest version (i.e., the Perkin-Elmer HS-40) allows the use of both the balanced pressure sampling technique with a single gas supply and the increased pressure sampling mode with two independent gas supplies. Both versions are shown in Figure 3-12 for working with open-tubular columns. The increased pressure sampling technique is particularly convenient if an open-tubular column is operated with a column inlet splitter as shown in Figure 3-12I, but it works also together with packed columns. The original balanced-pressure sampling technique is preferably used for splitless on-column injection (Figure 3-12II) onto a fused-silica, open-tubular column.

These systems require only that the original pressure in the vial that was built up during equilibration (p_v) be less than the headspace pressure, either the inlet pressure to the column p_i in the case of splitless injection or the increased pressure p_p in the case of split injection (cf. Figure 3-12).

$$p_i > p_v < p_p \qquad\qquad (3.3)$$

Otherwise, as soon as the sampling needle enters the vial, instead of pressurization in Step B (Figure 3-11B), an uncontrolled flow from the headspace toward the column will occur, leading to double peaking or peak splitting.

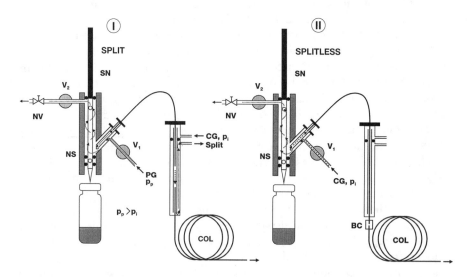

Figure 3-12. Modification of a balanced pressure system (see Figure 3-11) for (I) split operation with separate pressurization gas (*increased pressure sampling*) and for (II) splitless operation of open-tubular columns. CG = carrier gas, PG = pressurization gas, V = on/off solenoid valves, SN = movable sampling needle, NS = needle shaft; NV = needle valves; COL = column, p_i = carrier gas inlet pressure, p_p = pressurization gas pressure, BC = butt connector.

The pressure in most sample vials is usually generated from the saturation vapor pressure of water, not only in the case of aqueous samples, but even with many liquid or solid samples, due to their natural humidity. The saturation vapor pressure data for water at various temperatures compiled in Table 2-1 and plotted in Figure 2-2 should be consulted, whenever open-tubular columns are used for the analysis of aqueous or humid samples. This criterion may present a problem with columns that have a low pressure drop: there the inlet pressure p_i to the gas chromatograph may be less than the vial pressure p_V. In this case, by using the *increased pressure system* (Figure 3-12I) the vials can always be pressurized at a higher pressure p_p than the internal vial pressure. Another convenient possibility — preferably for short, wide-bore, open-tubular columns with a very small flow resistance — is to use an end restrictor (e.g., a 60 cm × 0.15 mm I.D. deactivated fused-silica capillary, connected to the column end by a butt connector). By varying the length of this restrictor capillary, any desired pressure drop can be adjusted.

As mentioned earlier, in the balanced pressure systems the volume of the headspace aliquot injected into the column is determined by the pressure and the time of transfer. Both are controlled very closely; thus, the injected gas volume can be accurately reproduced.

3.5.3 Pressure/Loop Systems

In pressure/loop systems, the sample vial is pressurized after equilibrium by the carrier gas to a preset value, just as in the balanced-pressure systems discussed in Section 3.5.2. However, in the next step the vial is opened temporarily toward the sample loop of a gas sampling valve and not directly to the column; thus now, the pressurized headspace gas from the vial will fill the sample loop. Next, the content of the loop is injected into the gas chromatograph in the usual way.

Figure 3-13 [13] illustrates the principles of an automated pressure/loop system and the three steps of sample handling. Figure 3-13A represents the setup during equilibration (standby). If needed, the sampling needle may also be purged in this step, by opening valve V_2. In Figure 3-13B ("pressurization" step), the sampling needle pierces the septum, and the pressurization gas flows through it into the headspace of the vial, until a preset pressure is reached. The third step consists of two parts: first (Figure 3-13C$_1$: "sampling") the pressurization gas flow is closed and the six-port valve is turned to a position in which the headspace of the vial is connected to the sample loop. The sample loop can be opened through the backpressure regulator (*BR*) to atmosphere, producing a flow of the gas from the vial (which is at the preset elevated pressure) through the loop toward atmosphere. The result of this would be that at the end, the headspace gas filling the loop is at atmospheric pressure. However, by the proper setting of the backpressure regulator, the final pressure in the loop can be set (e.g., to the carrier gas inlet pressure). Finally (Figure 3-13C$_2$: "sample injection"), the content of the sample loop is swept by the carrier gas into the separation column.

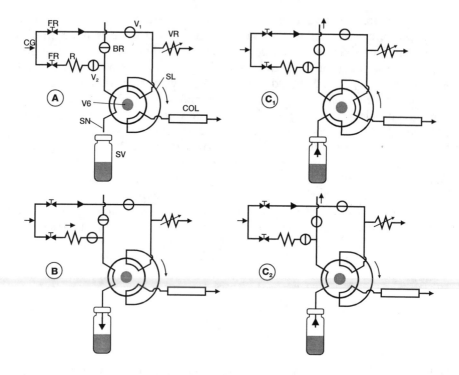

Figure 3-13. Principles of the pressure/loop system for headspace introduction into the gas chromatograph [13]: (A) equilibration (standby), (B) pressurization, (C₁) sampling (loop filling), (C₂) injection. CG = carrier gas, FR = flow/pressure regulator, R = restrictor, V = solenoid on/off valves, BR = backpressure regulator, VR = variable restrictor, $V6$ = six-port valve, SL = sample loop, SN = sampling needle, SV = sample vial, COL = column.

Naturally, the six-port valve as well as the transfer lines and the sampling needle are heated to the proper temperature to prevent any sample condensation.

By using an HS-GC system with a sample loop, one has a nominal "sample volume." However, this is only the geometric volume of the loop: what we want to determine is the actual amount of the analyte, expressed by the number of moles (n), which finally defines the analytical sensitivity. The number of moles present depends on the volume V_L, pressure p_L, and temperature T of the loop:

$$p_L \cdot V_L = n \cdot RT \qquad (3.4)$$

where R is the gas constant. Thus, specifying the loop's volume as the "sample volume" is meaningless. We shall deal with this question shortly (Section 3.5.6).

3.5.4 Conditions for Pressurization Systems

In the automated headspace systems based on pressurization, three parameters are set by the operator: pressures, times, and temperatures.

With regard to the pressurization time, a few seconds would be sufficient for just filling the vial up to the desired pressure. However, in practice, longer pressurization times (1–3 min) are recommended.* There are two reasons for this. The first has to do with the way the gas enters the vial through the needle: it is characterized by a laminar flow rather than immediate mixing by turbulence. Thus the incoming inert gas forms a gas bubble around the needle vent, causing a concentration gradient in the headspace of the vial. Diffusion in a gas is, however, fairly fast and thus, the gradient will soon disappear, resulting in a homogeneous distribution in the headspace. The longer pressurization time is needed to assure this homogenization.

There is controversy as to whether the concentration of the analyte in the gas phase is diluted when the headspace is pressurized with additional carrier gas. This depends on the dimensions in which the concentration is given. If expressed in molar concentration, there is indeed such a dilution effect, as described by the mole fraction $x_{G,i}$:

$$x_{G,i} = \frac{n_i}{n_{\text{total}}} \tag{2.21}$$

where n refers to the number of moles present in the headspace. However, the situation is different if $C_{G,i}$, the analyte concentration in the gas phase, is given in terms of weight/volume (i.e., in units of mass $W_{G,i}$ per volume V_G of the headspace):

$$C_{G,i} = \frac{W_{G,i}}{V_G} \tag{2.9}$$

Here the concentration is independent of the pressure because the volume of the vial will not change when it becomes pressurized. Thus, the resulting headspace sensitivity is also independent of the pressure in the vial.

The second reason for a somewhat longer pressurization time is that usually, the temperature of the incoming gas is different from the temperature of the vial's headspace, and some time is needed to achieve a uniform temperature in the vial.

The second variable set by the operator is the temperature of the needle, valve, and transfer line. This should be high enough to prevent condensation but not so high that the septum is burned with a too-hot needle. Usually a temperature at or slightly above the thermostatting temperature is recommended. We emphasize, as well, that in a heated transfer line the headspace gas is a mixture of air with trace concentrations of the analytes; thus too high a temperature may cause decomposition by oxidation.

---

* "Pressurization time" means the time the system is in the phase illustrated in Figure 3-11B; in the examples in this book, the pressurization time is 3 minutes unless otherwise mentioned.

3.5.5 Use of Open-Tubular Columns

The use of open-tubular (capillary) columns in HS-GC systems was first discussed in 1978–1979, by Kuck [14] and by Kolb, Ettre, and co-workers [15,16]. Today, following the general trend in gas chromatography, the use of this type of system is continuously increasing. Most of the examples shown in this book, therefore, utilized open-tubular columns.

A general question often asked is whether split or splitless injection should be used in capillary HS-GC. With the balanced-pressure sampling system, both techniques can be applied (cf. Figure 3-12). In the case of splitless on-column injection, the fused silica, open-tubular column either is threaded through the injector and the transfer line close to the sampling needle or is connected in the oven of the gas chromatograph by a butt connector to the inert fused-silica transfer capillary (see Figure 3-12II). Splitless on-column injection is necessary for the technique of cryofocusing, as discussed later (Section 3.8). In this case the carrier gas with column inlet pressure p_i is used to pressurize the vial, and this system is identical to the original principle of balanced pressure headspace sampling as shown in Figure 3-11.

With respect to split injection, let us not forget that in headspace sample transfer, an already homogeneous gas mixture is introduced into the gas chromatograph and thus, no vaporization followed by homogenization is necessary: we are mixing two homogeneous gases, the carrier gas and the effluent from the vial. In fact there is not much difference between split or splitless injection as far as the headspace sensitivity is concerned. Such a comparison is presented next (Section 3.5.6).

The admissible sample volume onto an open-tubular column is limited by the column's sample capacity. Overloading the capillary leads to band broadening, and this effect is independent of the type of sample introduction. In the case of the loop sampling systems, the volume of the headspace gas is controlled by the size of the loop and the loop flushing time, while with the balanced pressure sampling systems it is controlled by the injection (sample transfer) time. To avoid band broadening, the injection time should be limited to a few seconds. During this time the transferred gas volume is automatically set by the actual gas velocity. If we consider the same linear flow rate at the column inlet u_i, we can calculate the transferred headspace volume V_H from the injection time t and the cross section of the column Q_c:

$$V_H = Q_c u_i t \tag{3.5}$$

Table 3-6 gives values of the transferred headspace volume for three column diameters, assuming $t = 3$ seconds and $u_i = 30$ cm/s.

These data show that the sensitivity we get from these columns is proportional to its cross section Q_c. The effect of band broadening, on the other hand, depends on the sample transfer time and is thus independent of the column type. The injection time must be the same for all three columns. Therefore, a

Table 3-6 Injected headspace sample volume V_H as a function of the inner diameter and cross section CS of open-tubular columns at constant injection time t (3 s) and u_i (30 cm/s)

I.D. (mm)	CS (mm^2)	V_H (μL)
0.25	0.049	44.2
0.32	0.080	72.4
0.53	0.785	198.5

wide-bore capillary column with its higher gas flow is 4.5 times more "sensitive" than 0.25 mm I.D. narrow-bore capillary column, but of course the resolution is not as good. The actual volumes introduced into the columns are surprisingly small and, even in a wide-bore capillary column, the volume is only about 200 μL for the example, just given.

If a splitter is used (e.g., with a split ratio of 1:10, a 10 times larger gas volume is withdrawn from the vial during the same injection time, but is immediately divided 1:10 by the splitter), the sensitivity therefore should be the same. In the case of split injection as shown in Figure 3-12I, the headspace sample is further diluted in the injector by the column carrier gas, and it is only because of this effect that the sensitivity of split injection is usually about half that obtained with splitless injection. This dilution effect can be avoided with modern gas chromatographs having programmed pressure control because it is possible then to shut off automatically the carrier gas supply (CG in Figure 3.12I) or at least to reduce the column inlet pressure p_i during the sample transfer simultaneously with closing valve V_1. In this case there is practically no difference in the headspace sensitivity between split and splitless sample introduction, since the dilution effect in the injector is thus eliminated. Such a comparison is shown later (see Example 3.3).

Whether split or splitless sampling is preferred depends more on practical considerations and on the particular instrument design and less on the sample volume, which is in any case sufficiently large to allow withdrawal of a few milliliters of headspace gas with subsequent splitting. In general, it is more convenient to connect a headspace sampler to an already existing split/splitless injector and to operate it in the split mode. We note here, and discuss more fully in Section 3.8, that splitless on-column injection is mandatory only in the case of enrichment techniques by cryofocusing, where a splitter is no longer acceptable.

When the split ratio is high or injection times long (as in the case of cryofocusing), the actual volume will be less calculated from eq. 3.5 because the pressure in the vial (hence the velocity of the sample transfer) will decrease during sample transfer: in these cases eq. 3.5 is no longer valid.

3.5.6 Volume of the Headspace Gas Sample

Users often ask about the actual sample volume transferred from the headspace of the vial into the gas chromatograph for separation and analysis. There are two reasons for this question. The first is the general custom in analysis, where one always speaks about the amount or volume of the sample; thus, similarly, it is felt that knowledge of this property in HS-GC is also important. Here, however, the situation is different.

In headspace analysis, the expression "sample" has two meanings. First, it means the original (liquid or solid) sample placed into the headspace vial. Knowledge of its volume V_S is important, since it influences the volume of headspace gas V_G and the phase ratio β:

$$V_V = V_S + V_G \tag{2.1}$$

$$\beta = V_G/V_S \tag{2.2}$$

(V_V = volume of the sample vial) and the phase ratio is an important value that is involved in quantitative calculations. In general, V_S is known: after all, it is the volume of the sample the user placed into the vial.

For analysis, an aliquot of the headspace gas is transferred into the gas chromatograph, and this sometimes is also called the "sample." We prefer the expression "headspace sample volume (V_H)." It is important to realize that this volume is not included in quantitative calculations: thus, the analyst does not need to know the exact volume of this transferred aliquot. What is needed is the possibility of exact sample-to-sample *reproducibility* of the transferred gas volume and the flexibility of changing it.

There is, however, another reason for sometimes wanting to know the volume of the transferred gas: to establish comparative conditions with the various headspace systems. This question may be raised because—for historical reasons—many standard HS-GC methods were originally developed using a manually operated gas syringe and the "sample volume" (i.e., the nominal volume of the syringe) is indicated in them. This is a difficult question, and one must consider whether it refers to the nominal volume of the syringe or the actual volume of the aliquot taken from the vial. In other words, the fundamental issue is the meaning of "volume."

The ASTM standard practice dealing with the determination of volatiles in polymers [5] suggests the manual withdrawal of 1.0 mL headspace gas V_H with a gas syringe, so let us start with this volume. As discussed earlier, however, the headspace gas we sample from the vial is under pressure generated by the elevated temperature of thermostatting. If we have an aqueous sample and thermostat it at 80 °C, the internal *absolute* pressure in the vial will be 148.5 kPa* because the saturation pressure of water at this temperature is 47.2 kPa. Upon withdrawal of an aliquot of the headspace gas with a standard

---

* Again, we neglect the possible contribution of the partial pressures of other components present in the sample.

syringe, this pressure extends into the volume of the syringe barrel; however, when the syringe is taken out of the vial, its needle is open to atmosphere and thus, the withdrawn gas will expand, loosing almost half of the withdrawn sample. This sequence can be calculated using the basic pressure–volume relationship:

$$p_1 \cdot V_1 = p_2 \cdot V_2 \tag{3.6}$$

In the present case, $p_1 = 101.3\,kPa$, $V_1 = 1.0\,mL$ (conditions in the syringe), and $p_2 = 148.5\,kPa$ (absolute pressure in the vial); thus V_2 (the corresponding headspace gas volume) is $(101.3 \times 1.0)/148.5 = 0.68\,mL$. In other words, the volume of the aliquot taken from the vial was actually only $0.68\,mL$, although we treat it as $1\,mL$.

On the other hand, if a pressure-lock type of syringe is used, the pressure of the withdrawn sample is preserved: in this case, we withdrew $1\,mL$ of the headspace gas at headspace pressure. In other words, just selecting the type of syringe produces a significant difference in the volume of the sample introduced into the gas chromatograph. Other variants are atmospheric pressure (which is changing) and vial pressure. In essence, what we stated here also refers to the automated syringe systems, except that, as stated earlier, their syringe is open to the atmosphere.

In the case of *loop injection*, the declared volume is the geometric volume of the loop. If the loop is open to atmosphere, the situation is the same as just given for a syringe: when the pressurized gas expands through the loop to atmosphere, the pressure will decrease exponentially, and the amount of the analyte, expressed as the number of moles in the fixed volume, will correspondingly decrease during this process. If the pressure changes during the expansion of the pressurized headspace gas from the original vial pressure down to atmospheric pressure, the number of moles in the loop will steadily decrease. Thus, the amount of analyte in the loop depends on the headspace pressure in the vial and the time between the start of sample transfer into the loop and the instant when the valve is turned to the injection position (cf. Figure 3-13).

The other possibility is to keep the gas in the loop under a certain pressure with the help of a backpressure regulator. In this case, the amount of analyte will be different again, but can be calculated using eq. 3.6, assuming, naturally, that the pressures in the vial and in the loop are known.

If a different gas volume is to be injected, then the loop (i.e., its geometric volume) and/or the pressure within it must be changed. If, however, the content of the loop is only partially transferred onto the column by stopping the flushing period (loop flushing time), any calculation of the volume will become even more complicated.

In the case of *balanced pressure systems*, the actual sample volume transferred into the column depends only on the carrier gas velocity at the column inlet and the time of transfer (which can be accurately controlled, assuring the exact reproducibility of the transferred volume) but is independent of the pressure. If needed, this volume can be calculated from the available data in the following way. First we calculate F_i, the carrier gas flow rate at column inlet, knowing F_a, the flow rate at column outlet and the (absolute) column inlet and (atmospheric) outlet pressures (p_i and p_a), based on the basic pressure–volume relationship (eq. 3.4). Usually F_a is measured anyway in a GC measurement, while p_i is a set value,* thus these are readily available values:

$$F_i = \frac{p_a}{p_i} \cdot F_a \tag{3.7}$$

Assuming that $p_i = p_V$, this will also be the flow rate of the headspace gas from the vial into the column. Hence, the gas volume transferred in time t will be:

$$V_{gas} = F_i \cdot t \tag{3.8}$$

We present next an example illustrating how, from readily measured values, the actual volume introduced from the headspace into an open-tubular column, and from it the corresponding amount of the analyte, can be established. The analyte is 1-bromo-4-fluorobenzene (BFB), the specified compound for the calibration of mass spectrometers. First we deal with the case of *splitless sampling.*

Example 3.2

We used an open-tubular column (25 m × 0.32 mm I.D., coated with methyl silicone stationary phase; film thickness: 0.1 μm) at 60 °C. The standard solution of BFB was prepared using acetone as the solvent; the concentration of the solvent was 1 μg/μL, and 2 μL of this solution was introduced into the vial. The following conditions were used:

　column outlet (atmospheric) pressure: $p_a = 96.97\,kPa$

　column head pressure: $\Delta p = 109.94\,kPa$

　column (abs) inlet pressure: $p_i = \Delta p + p_a = 109.94 + 96.97 = 206.91\,kPa$

　measured flow rate at column outlet: $F_a = 4.5\,mL/min$

　vial temperature (120 °C): $T_v = 393.16\,K$

Since F_a was measured with a bubble flow meter at 22 °C ambient temperature, corrections for dry gas conditions and vial temperature are needed to ensure more accurate results:

$$F_{c,o} = F_a \cdot \frac{T_v}{T_a} \cdot \frac{p_a - p_w}{p_a} \tag{3.9}$$

where T_a is the ambient temperature ($= 295.16\,K$), T_v is the vial temperature ($= 393.16\,K$), $p_a = 96.97\,kPa$, and p_w (the partial vapor pressure of water at ambient temperature) is 2.637 kPa. The result of the calculation is $F_{c,o} = 5.83\,mL/min$.

---

* One has to be careful with "inlet pressure." Colloquially, one usually speaks about the "inlet pressure," although the term really means the pressure drop Δp (i.e., the gauge pressure). For p_i we need the absolute inlet pressure ($p_i = \Delta p + p_a$).

Next, we calculate the flow rate at the conditions of the vial (i.e., at column inlet pressure), using eq. 3.7, but replacing F_a by $F_{c,o}$:

$$F_i = \frac{p_a}{p_i} \cdot F_{c,o} = \frac{96.97}{206.91} \cdot 5.83 = 2.73 \, \text{mL/min} \tag{3.7a}$$

Injection time (sample transfer time) was $t = 0.05$ minutes. Thus, the volume of the transferred headspace gas is (eq. 3.8):

$$V_{gas} = 2.73 \times 0.05 = 0.137 \, \text{mL}$$

The concentration of BFB in the vial ($V_V = 22.3 \, \text{mL}$) was $2 \, \mu\text{g}/22.3 \, \text{mL} = 89.7 \, \text{ng/mL}$. Thus, the amount of BFB in $137 \, \mu\text{L}$ gas was $89.7 \times 0.137 = 12.28 \, \text{ng}$. The peak area obtained for this amount was 12,410 counts.

If split sampling is used instead of splitless, the calculation is in principle the same. The actual amount introduced into the column depends on the linear flow rate at column inlet and thus, the inlet pressure, and the injection time. These parameters are not altered by opening the splitter. Only if we want to determine the total volume that is withdrawn from the vial do we need to include the split ratio in the calculation. These considerations, however, are true only if during the sample transfer the carrier gas supply (CG in Figure 3-12I) to the GC-injector is shut off (e.g., by programmed pressure control) simultaneously with valve V_1, to avoid the dilution effect in the injector, discussed in Section 3.5.5. This technique was used in the following example.

Example 3.3

We first carry out a splitless injection to obtain a peak area/amount calibration factor. In this particular case, it was 12,410 counts for 12.28 ng (see Example 3.2). Next, we carry out a split injection under the desired conditions, using a split ratio 1:20. In our test, we obtained a peak area of 13,330 counts. Thus, the corresponding amount entering the column was

$$\frac{13,330}{12,470} \cdot 12.28 = 13.13 \, \text{ng}$$

This example proves the earlier statement that in principle there is no difference between split (11.13 ng) and splitless (12.28 ng) headspace sampling as far as the transferred analyte amount and thus the sensitivity are concerned. The 6.5% difference may be explained by a faster sample transfer through the transfer line in the case of split injection, which results in a slightly higher injected sample amount during the same injection time. In the splitless injection we transferred $137 \, \mu\text{L}$; considering the split ratio of 1:20, the total gas volume taken out of the vial was 2.9 mL.

This calculation assumes that during sample transfer the flow rate from the vial to the column remains constant and is equal to the column flow rate. However, the vial is a limited reservoir of gas, and by expansion of the pressurized gas either onto the column or through a loop, the pressure in it—and thus the flow rate from it—decreases exponentially. According to

Hinshaw and Seferovic [17], the deviation from the linear model is below 1% for a 0.25 mm I.D. column, and even for a 0.53 mm I.D. column the difference due to the exponential decay in the flow is only a few percent and thus negligible with the normal injection times of a few seconds. Such calculations should therefore be made up with short injections times only where the flow rate remains practically constant. Even when a small split ratio is selected, the difference from splitless injection is negligible, as illustrated by the preceding practical example.

However, in the case of packed columns, or with open-tubular columns operated at a high split ratio, the relationship of peak area to transfer time may become slightly nonlinear. Such a nonlinearity is also found if an open-tubular column is operated with long injection times (up to several minutes when using cryofocusing). This is shown in Figure 3-14 for the example of toluene determined from an aqueous solution with cryofocusing and injection times up to 5 minutes.

Finally it is emphasized that the determination of the absolute analyte amount or the true sample volume is necessary only if it is required to

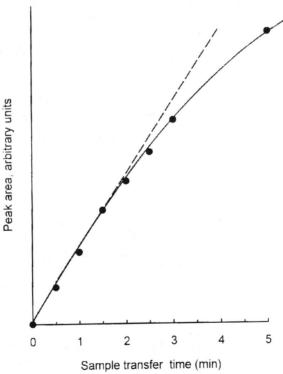

Figure 3-14. Increase of the peak area values of 0.3 ppm toluene in water at 80 °C with increasing sample transfer times (minutes) using cryogenic focusing. Splitless sampling onto a 0.32 mm I.D. fused-silica, open-tubular column.

determine absolute data such as the detector sensitivity. (For more discussion, see Chapter 9). For analytical applications of HS-GC, these values are meaningless and of no practical consequence, since the actual volume of the introduced sample is not the most important criterion and in general, it is not involved in quantitative calculations: what is important is the fact that it is exactly reproduced. The admissible volume is restricted by the beginning of peak broadening by overloading the column, and thus the deciding criterion is maintaining the chromatographic resolution. Success in this regard depends on many parameters, such as the type of the column, its diameter, the flow rate, the loop flushing time in the case of loop sampling or the injection time in the case of balanced-pressure sampling, and also whether the focusing effect of temperature programming is utilized. All these parameters still must be empirically optimized.

3.6 Instrumentation of MHE

We have already discussed in detail the theoretical background of multiple headspace extraction (Section 2.6); further questions associated with its use in quantitative analysis will be detailed in Section 5.5. As explained, MHE is a continuous gas extraction carried out stepwise, with sequential analysis of a number of headspace samples taken from the same vial. After each analysis, the pressure in the vial is reduced to atmospheric and reequilibrated before the next analysis.

As demonstrated in connection with Figures 3-11 and 3-13, in automated headspace–gas chromatography utilizing pressurization, the analysis consists of three steps. In Step A, the vial is thermostatted until it has reached its equilibrium ("equilibration"). During this step the carrier gas is conducted directly to the column, in the usual way. After equilibrium has been reached, the vial is pressurized by the carrier gas either to the column inlet pressure or to a preselected pressure (Step B). In the third step (Step C) an aliquot of the headspace gas is introduced into the column, either directly or via a fixed sample loop. After the sample transfer, the system is ready for the next headspace vial.

In multiple headspace extraction, two additional steps are added to the three just listed. First, the pressure in the vial is released to atmosphere ("venting"). This may be done manually, by puncturing the septum with a syringe needle, or by the proper construction of the system. After the pressure release, the vial must be reequilibrated for the next headspace analysis.

In modern HS-GC systems these additional steps are also carried out automatically. We illustrate them here with the example of the balanced pressure system. For venting, the sample needle is moved slightly downward, so that the upper opening of the hollow part is now open to atmosphere (Step D, Figure 3-15D). For reequilibration, the sampling needle remains in the vial but

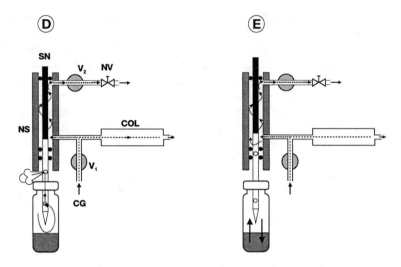

Figure 3-15. The two additional steps of operation in the balanced pressure system for multiple headspace extraction (MHE): (D) venting, (E) reequilibration (standby). CG = carrier gas, V = on/off valves, SN = movable sampling needle, NV = needle valve, NS = needle shaft, COL = column.

it now moved upward so that the upper opening of the hollow part is now in a closed compartment between two O-rings in the lower part of the needle shaft (Step E, Figure 3-15E): in this way, the headspace will remain sealed. In other words, the needle remains in the vial during the whole operation and the vial's septum is pierced only once, at the beginning of the multistep procedure, avoiding any leakage that may result from multiple piercing of the septum. The next analysis begins with again pressurizing the vial (Step B, Figure 3-11B)

3.7 Backflushing

Backflushing is a well-established method in gas chromatography, and it has particular importance in headspace analysis for two reasons. In general, we are interested in the more volatile sample components, and thus elimination of the heavier end can speed up the analysis. Also, when solid samples are dissolved for headspace analysis, usually the solvent has a higher boiling point than the volatile compounds of interest; in addition, when a modifier (displacer; see Section 5.6), is used, the solvent also has a higher boiling point. Thus, the elution of the vapors of the solvent or the modifier through the full length of the separation column would unnecessarily prolong the analysis time.

With a packed column, even in self-built systems, backflushing of the column can be accomplished by using a heated six-port valve. The ASTM

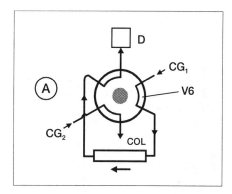 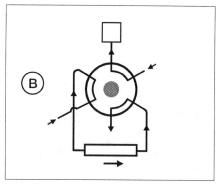

Figure 3-16. Backflushing a column [5]: (A) analysis position, (B) column backflush. $V6$ = six-port valve, COL = column, D = detector; CG_1 = main carrier gas flow (from the injector); CG_2 = auxiliary carrier gas flow.

standard practice discussing the HS-GC method for the analysis of volatiles present in polymers [5] describes such a system; its functional schematic is shown in Figure 3-16.

In the balanced pressure system, the need for a multiport valve is eliminated [10]. It was also recognized that the whole column does not have to be backflushed: when the first volatile analytes emerge from the column, the heavier sample components are still in the first part of the column, therefore, it is enough to backflush its front while containing the regular carrier gas flow ("foreflush") through the second part of the column. For this arrangement, two identical columns are used in series (Figure 3-17), and the auxiliary carrier gas flow is brought to the middle point between the two columns and its pressure adjusted to a value that is about half of the column inlet pressure. This auxiliary carrier gas flow is permanently open. This two-column arrangement has the additional advantage that the column bleeding to the detector is not interrupted in the backflush position (F_2), and therefore no baseline upset is observed (see Figure 3-18), not even at high detector sensitivity (see Figure 3-19). In the regular operation (F_1) the two columns are in series. When switching to backflushing (F_2), the main carrier gas flow is disconnected by means of valve V_1. Simultaneously, the sampling needle SN moves a small step down from its closed compartment until its lower vent is open to the atmosphere through which column C_1 is now backflushed. At the same time, column C_2 continues to receive the carrier gas flow in the usual way: thus, the volatile analytes that were in the column at the moment of switching continue to elute and appear in the chromatogram.

In the case of open-tubular (capillary) columns, the system must be modified because of the usual problems associated with connecting two capillary columns in series. The modification consists of using a special middle piece [18] (see inset in Figure 3-17), which may be constructed from a Swagelok

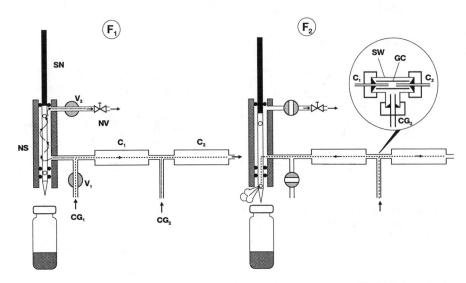

Figure 3-17. Column backflushing in the balanced pressure system [10]: (F_1) analysis position ("foreflushing"), (F_2) backflushing the first and foreflushing the second column. CG_1 = main carrier gas flow, CG_2 = auxiliary carrier gas flow, V = on/off solenoid valves, SN = movable sampling needle, NV = needle valve, NS = needle shaft, C_1, C_2 = columns. Inset: SW = Swagelok T-fitting, GC = glass capillary tube.

T-piece in which a short glass capillary (about $20 \, mm \times 0.40 \, mm$ ID* for $0.32 \, mm$ I.D. fused-silica capillary columns) is placed. Both columns C_1 and C_2 are inserted at each end of this tube until they almost meet in the middle. At this position the two columns are tightened onto the Swagelok T using graphite or Vespel ferrules. The auxiliary carrier gas flow enters through the third arm of the Swagelok T and flows continuously along the glass insert, thus flushing any dead volume.

It is advisable to measure the pressure at the point between the capillaries. This can be done by turning the pressure gauge to the auxiliary gas line fully counterclockwise. The actual pressure in the auxiliary gas line should then be adjusted to a value about $20 \, kPa$ higher than the measured pressure. With modern programmed, pressure-controlled instruments, this pressure can be adjusted even more precisely. This will assure that the connecting device is continuously swept in position F_1.

Figure 3-18 shows an example for the use of backflushing with open-tubular columns. When analyzing the BTEX content (benzene, toluene, ethylbenzene, and the xylenes) of a used engine oil, a number of volatile hydrocarbons emerging after the xylenes prolong the analysis and make the chromatogram more complicated. Switching to backflushing position at 15 minutes, just

---

* The internal diameter of this piece should be just slightly larger than the outside diameter of the two fused-silica columns C_1 and C_2.

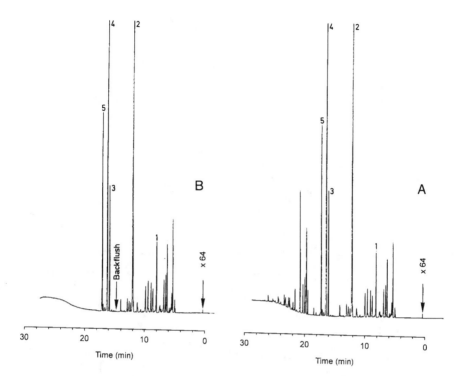

Figure 3-18. Determination of the aromatic hydrocarbon (BTEX) content of used engine oil, using an open-tubular column (A) without backflush and (B) with backflush. *HS conditions*: Sample volume: 1 mL; equilibration at 80 °C. *GC conditions*: Columns: Two 25 m × 0.32 mm I.D. fused-silica, open-tubular in series, coated with methyl silicone stationary phase. Column temperature: isothermal at 40 °C for 8 minutes, then programmed at 8 °C/min to 160 °C. Carrier gas: Helium; main flow: 190 kPa (0.7 mL/min); auxiliary carrier gas: 120 kPa. Backflush on: 15 minutes. Split sampling, split 1/57. Flame-ionization detector. *Peaks*: 1 = benzene, 2 = toluene, 3 = ethylbenzene, 4 = m/p-xylene, 5 = o-xylene.

before the emergence of ethylbenzene, will continue the elution of ethylbenzene and the xylenes (which were already in the second column when backflushing started), while backflushing the higher hydrocarbons that were still in the first column at the instant of switching. In a second example (Figure 3-19), we have the analysis of some solvents, at a concentration of 10 ppm each, dissolved in dimethylformamide. This sample was used as an external standard for the determination of organic volatile impurities (OVIs) in pharmaceutical drug samples utilizing the solution approach (see Section 4.2) according to the U.S. Pharmacopoeia [19]. The concentration of 10 ppm of these solutes corresponds to 100 ppm in the original drug sample, if prepared as a 10% solution in DMF. By backflushing, the DMF peak could be eliminated and the analysis time shortened.

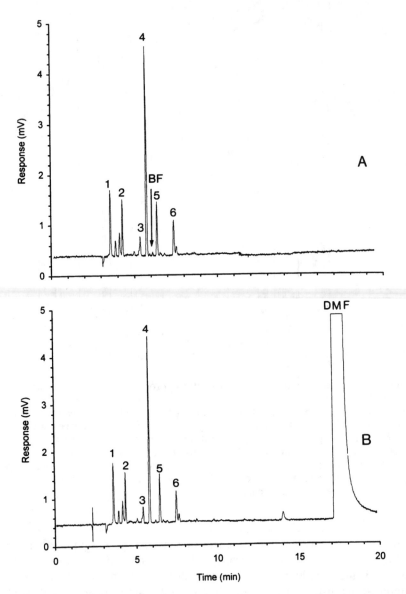

Figure 3-19. Analysis of solvents dissolved in dimethylformamide (DMF) at a concentration of 10 ppm each, used as an external calibration standard for the determination of organic volatile impurities (OVIs) in pharmaceuticals (A) with and (B) without backflush. *HS conditions*: 1 mL solution in DMF. Equilibration at 80 °C for 45 minutes. *GC conditions*: Columns: Two 25 m × 0.32 mm I.D. fused-silica, open-tubular in series, coated with bonded cyanopropyl (14%) methyl silicone (OV-1701); film thickness: 1 μm. Column temperature: 80 °C isothermal. Carrier gas: Helium 150/100 kPa, split sampling, split flow 20 mL/min; auxiliary carrier gas: 85 kPa. Backflush (*BF*) on at 6 minutes, off at 12 minutes. Flame-ionization detector. *Peaks*: 1 = methanol, 2 = dichloromethane, 3 = chloroform, 4 = benzene, 5 = trichloroethylene, 6 = dioxane.

3.8 Cryogenic Sample Concentration

As we have already demonstrated, HS-GC is a method for trace analysis par excellence: when the proper conditions are selected, analytes present in very low concentrations in the original sample can be enriched in the gas phase.

Even with this enrichment, however, the concentrations in the headspace are sometimes very low, either because the analytes are already present in very low concentrations in the original sample or because the analytes of interest have a relatively low vapor pressure at the temperature of equilibration. Also if we want to use a detector with a low sensitivity, such as Fourier transform infrared (FTIR) for peak identification, a large gas sample volume will have to be introduced into the column. The admissible volume of a gas sample depends on the type of column. As we have already seen, a packed column accepts a larger gas volume compared to an open-tubular column and of these, a wide-bore capillary column has a higher capacity than a narrow-bore column (cf. Table 3-6). Here we have the classical conflict in gas chromatography between sensitivity (i.e., sample capacity) and resolution. On the other hand, the available volume of the gas sample in the headspace vial is large enough for better sensitivity: from the headspace vial, only a few percent of its volume is transferred into a capillary column in the normal analysis (e.g., in the case of a 20 mL vial, a 200 μL headspace sample represents only 1%).

The admissible volume of a gas sample onto a capillary column is restricted by band broadening and deterioration of resolution. The solution to this problem is simple: one starts by separating the bulk of the air in the headspace sample from the volatile analytes present. This can be done either by an adsorption step as in the dynamic headspace techniques, or condensation in a cold trap. In both cases the air passes through the trap while the analytes are retained. Adsorption is preferred if large gas volumes are to be handled, as in the case of purge-and-trap techniques, while the smaller volumes present in static HS-GC are easier to handle by cold trapping.

The terms "cold trapping" and "cryofocusing" are often used synonymously, but they should be differentiated. *Cold trapping* refers to a cold spot, with a constant low temperature, while *cryofocusing* includes a temperature gradient along a cryogenic trap where an additional band focusing effect is achieved.

3.8.1 The Need for Cryogenic Sample Concentration

We shall discuss the effects of sample volume, temperature, and band broadening using as an example the headspace analysis of 2 g of ground cheese samples with splitless injection and two sampling times (Figure 3-20) [20]. In the first case, sample transfer time was 4.8 seconds,* the longest time generally recommended for such open-tubular columns. Under these conditions, sharp

---

* In modern computer-controlled instruments, times are set as fractional minutes; thus, a sample transfer time of 4.8 seconds corresponds to 0.08 minute and 24 seconds to 0.4 minute.

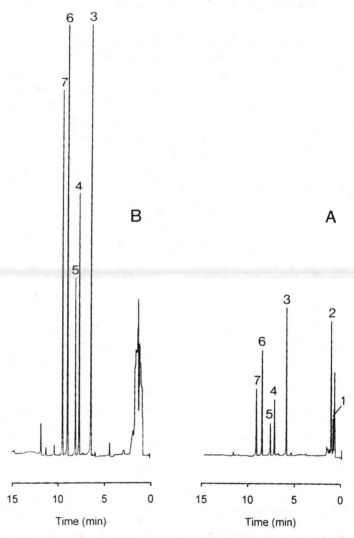

Figure 3-20. Headspace analysis of a cheese sample with different transfer times [20]. *HS conditions*: Sample: 2 g ground cheese, equilibrated at 90 °C for 60 minutes. Transfer time: (A) 4.8 seconds and (B) 24 seconds. *GC conditions*: Column: 25 m × 0.32 mm I.D. open-tubular, coated with free fatty acid phase (FFAP); film thickness: 1 μm. Column temperature: programmed at 8 °C/min from 70 °C to 180 °C. Splitless injection. Flame-ionization detector. *Peaks*: 1 = acetaldehyde, 2 = ethanol, 3 = acetic acid, 4 = propanoic acid (130 ppm), 5 = isobutanoic acid, 6 = *n*-butanoic acid, 7 = isopentanoic acid (85 ppm).

peaks were obtained, but the peaks were very small. However, when sample transfer time was increased by a factor of five, to 24 seconds, a significant deterioration in the shape of the early peaks could be observed. This is due to overloading caused by the long transfer time and the long initial band entering the column. Later peaks are still sharp because of the usual focusing effect of the stationary phase at the initial lower temperature of the temperature program. Thus, these later peaks, increased in height proportionally to the sample size increase.

This problem of undesired peak broadening or distortion due to a too-large sample volume can be solved by the technique of *cryogenic trapping* (*cryofocusing*).

In cryogenic trapping the carrier gas flow containing the broad sample band is conducted through a cold trap where the analytes condense or are retarded. Next, by rapid heating up of the trap, the accumulated analytes are released, but now in the form of a sharp band, which is then carried by the carrier gas onto the column. Cryogenic trapping is a common technique in gas chromatography, particularly in combination with sampling techniques, either when large gas volumes have to be injected into a low capacity, open-tubular column or when the sample introduction period is too long for the requirements of an instantaneous sample introduction into the column. Cryogenic trapping is even frequently used for an additional band focusing in dynamic headspace analysis ("purge-and-trap"), if the thermal desorption does not deliver the released analytes sufficiently fast; however, it is an important technique in static headspace measurements, further extending the detection limits to very small concentrations.

3.8.2 Systems for Cryogenic Trapping

There are essentially four ways to carry out cryogenic trapping (cryofocusing). In the first case, a small trap is used before the column, which is cooled by a separate gas flow, conducted through a Dewar flask containing liquid nitrogen. This type of system was described by Kuck [14], who was the first to use cryogenic sample concentration in HS-GC with glass open-tubular columns. His trap was a 3 cm long section of the glass capillary column, about 3 cm after the column head. In this way, using splitless sample injection and sample transfer times of a few minutes, he could achieve an enrichment of a factor of 40, using the balanced-pressure HS-GC system.

In the second variant of cryogenic trapping in HS sampling, the whole GC oven, with the column in it, is initially cooled by means of liquid nitrogen or CO_2. Such subambient operation has been in use in gas chromatography for a long time — since the start of temperature programming, when analyzing wide-boiling-range samples also containing low-boiling compounds. Its use for dynamic headspace analysis was first described by Pankow and Rosen [21], who called it the "whole-column cryotrapping (WCC)" procedure. This technique was applied by Wylie [22] in static HS-GC, for the analysis of complex natural samples. Using the pressure/loop HS system, he developed the so-called multiple headspace injection (MHI) technique, in which the analyst makes several rapid injections from the sample (placed in a number of vials) into the same column held at cryogenic temperature: these multiple samples are all cryotrapped as a plug at the column head and eluted jointly by the temperature-programming of the column. Figure 3-21 illustrates Wylie's results by showing the HS analysis of a cola-type drink, with a single injection and with three injections using the MHI procedure. This technique, but with multiple injections from the same vial, is standard procedure with some commercial pressure/loop systems (e.g., the Tekmar Model 7000/7050 headspace analyzer).

The technique of "whole-column cryotrapping," where the oven of the gas chromatograph including the column is cooled down, is necessary if the chromatographic separation needs subambient temperature (e.g., as in the case of gas analysis). On the other hand, the problems associated with this technique are the high consumption of the coolant and the long time needed at the end of an analysis to cool the whole oven to cryogenic temperature again for the next analysis. If the task is only to trap the sample at the head of the

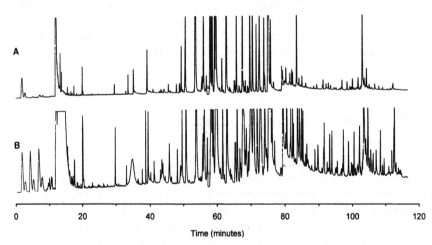

Figure 3-21. Analysis of the headspace of a cola-type soft drink, with cryotrapping: A = single injection of 1 mL headspace, β = three injections using the MHI procedure. *HS conditions*: Sample volume: 10 mL to which 6.0 g anhydrous Na_2SO_4 was added. Vial volume: 22 mL. Equilibration at 90 °C for 45 minutes. Sample loop: 1 mL. *GC conditions*: Column: 50 m × 0.32 mm I.D. fused-silica, open-tubular, coated with bonded methyl silicone stationary phase; film thickness: 0.52 μm. During the injection of headspace the GC oven was at −50 °C/min to 10 °C; after a 5 minute isothermal period at 10 °C, it was programmed at 1.5 °C/min into 215 °C. Split injection, split 1/36. Flame-ionization detector. Carrier gas: Helium, average velocity: u = 29 cm/s.
Source: Reproduced from ref. 22 with permission of P. L. Wylie and *Chromatographia*.

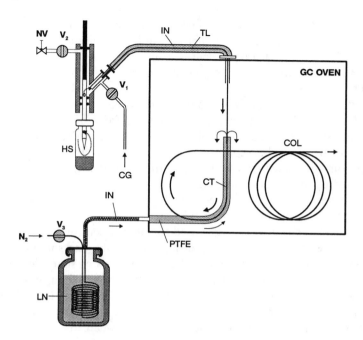

Figure 3-22. On-column cryofocusing at the front of the open-tubular column in the balanced pressure system [20]. HS = headspace sampling position (see Figures 3-11 and 3-12II), CG = carrier gas, V = solenoid valves, LN = liquid nitrogen bath, IN = insulation, TL = heated transfer line containing a fused-silica capillary, COL = analytical open-tubular column, CT = cryo-trap, first coil of the column (about 60 cm), $PTFE$ = Teflon tube.

column, then it is sufficient to only cool that part of the column. This "on-column head trapping" was first applied by Kuck [14] and later by Takeoka and Jennings [23], who formed the front of the fused-silica column (about 25 cm) into a U-shape and inserted it into a Dewar flask, filled with liquid nitrogen. A 500-μL headspace gas sample was injected with a gas-tight syringe, and 30 seconds after the injection was completed, the Dewar was removed and column temperature programming commenced.

An advanced version of on-column head trapping is where the front of the column is cooled dynamically [20]: Figure 3-22 outlines this system. Half of the first coil of the fused-silica capillary column or a transfer capillary is jacketed by a Teflon tube, and nitrogen gas, precooled in a liquid nitrogen (or CO_2) bath, is conducted through this tubing, in the opposite direction to the carrier gas flow inside the fused silica capillary column. In this way, a strong temperature gradient is generated, with the temperature at the entrance of the cryotrap near to the temperature of the cooling bath, while at the other end, the temperature can be adjusted by the flow rate of the cooling gas, usually to temperatures of -30 to $-10\,°C$.* If desired, this temperature at the end of the cryotrap can be measured by a thermocouple and displayed on a readout. Because of this temperature gradient, the operation using the setup shown in Figure 3-22 can truly be called cryogenic *focusing*, as opposed to trapping in the other variants.

The flow of the cooling gas should be switched on a few minutes before pressurization and sample transfer, to cool the front of the column. When sample transfer is finished, the cooling gas is switched off: the cooled part of the column will now be rapidly heated internally by the warm carrier gas coming from the heated transfer line, and this heating is much faster than can be achieved by any external heating.

A few general comments are necessary in connection with cryogenic sample trapping (cryofocusing) using open-tubular columns.

In general, cryogenic trapping is used to enhance sensitivity, permitting the analysis of ultratrace sample components. Obviously, one should avoid any unnecessary sample loss, such as would be represented by split injection. Thus, in trace analysis with open-tubular columns and using cryofocusing enrichment techniques, splitless on-column injection must always be used. The term "splitless on-column injection" is selected here to avoid confusion with the splitless injection technique with liquid samples, where the outlet splitter is closed before a liquid sample is injected and opened again soon after. In this case the resulting focusing effect is achieved by the "solvent effect." Using cryo-HS-GC, the headspace sample should be transferred directly into the open-tubular column without passing the unnecessary volume of the injector.

A further question is the need for flexibility in selecting the headspace volume to be transferred. In the balanced-pressure systems one can select practically any volume from a single vial by adjusting the sample transfer time, from a few seconds up to minutes, and this is highly reproducible. Figure 3-23 compares a regular headspace analysis using a sampling time of 3.6 seconds with cryofocusing and 30 seconds transfer time. We repeated this measurement nine times with good precision: the relative standard deviation of the highest peak in chromatogram B was 1.5%. With the pressure/loop systems, larger sample volumes can be injected only with help of the multiple headspace injection techniques, by repetitive injection from a single vial with accompanying multiple piercing of the septum, or from several vials into a cooled column.

In cryogenic trapping at the head of the column, the selection of the stationary phase needs special consideration. In the case of silicones, the phase behaves as a true stationary phase even at low temperatures: for example, the glass transition point of dimethylsilicone is $-114\,°C$. Therefore, in such a column the "trapping" of the analytes is due not to freezing but to real chromatographic retention: at the very low temperature of the trap, the

---

* Obviously these temperatures represent the temperature of the cooling gas in the Teflon tube, not necessarily inside the capillary column.

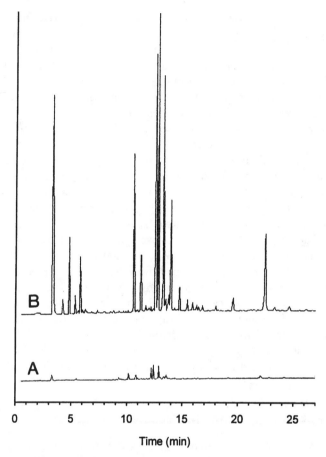

Time (min)

Figure 3-23. Analysis of volatile aroma compounds in a scented detergent, by HS-GC: A = normal headspace analysis, 3.6-second transfer time; B = cryofocusing in the system shown in Figure 3-22, with 30-second transfer time. *HS conditions*: Sample: 1 g solid detergent, equilibrated at 90 °C for 60 minutes. *GC conditions*: Column: 50 m × 0.32 mm I.D. fused-silica, open-tubular, coated with bonded phenyl (5%) methyl silicone stationary phase; film thickness: 5 μm. Column temperature: (*A*) programmed at 3 °C/min from 60 °C to 200 °C, (*B*) programmed at 20 °C/min from 40 °C to 60 °C and then programmed at 3 °C/min to 200 °C. Splitless injection. Flame-ionization detector.

partition coefficient K is very high, and thus its retention factor k will also be very high to fulfill the fundamental relationship

$$K = k \cdot \beta \tag{3.10}$$

because the phase ratio β of a column is a fixed value. This point has been discussed by Takeoka and Jennings [23].

If a polar stationary phase is used, even its melting point might be higher than the trapping temperature: thus the phase becomes solid and crystalline at the low temperatures of the cryotrap. In such a case analyte immobilization would be due to a combination of freezing and surface adsorption on the stationary phase, which is not a well-defined process. Therefore, in such a case it is advisable to use a short piece of an open-tubular column coated with a bonded silicone phase in the trap, which is then connected in series to the analytical column in the usual way used in capillary gas chromatography. Such examples are shown later (Figures 3-27 and 5-24).

Samples with an excess of a polar solvent are sometimes better trapped in a fused-silica capillary that was not deactivated (i.e., still has an active hydrophilic surface). Examples are aqueous solutions or alcoholic beverages in which the main compound—either water or ethanol—is not dissolved in a hydrophobic nonpolar stationary phase such as dimethylsilicone, at such low temperatures. If not dissolved, these are likely to form droplets that pass the cryotrap unretained, leading to double peaks, peak splitting, and/or peak distortion. The same effect may be observed if a deactivated fused-silica capillary with a hydrophobic surface is used in the cryotrap. With an active fused-silica capillary in the cryotrap, however, these polar compounds are spread and frozen smoothly along the hydrophilic surface, thus forming a temporarily stable "stationary phase" (liquid and/or solid), with no risk of plugging the capillary. Thus, the cryotrap may be considered to be a "retention gap" at lower temperatures. A hydrophilic surface inside the fused-silica capillary can also be achieved with a polar-deactivated, fused-silica capillary. Such a cryotrap was used in an example shown later (Figure 3-32).

3.8.3 Influence of Water in Cryogenic HS-GC

If the sample contains water—either as the solvent or present as humidity—then the water content of the headspace will also be trapped: in fact, it may even block the capillary column by ice formation. Other detrimental effects of the trapped water are peak distortion (particularly with early peaks), which is somewhat comparable to the "solvent flooding" effects in cold splitless injection into open-tubular columns, and shifts in the retention times, as described by K. Grob and Burns et al. [24–26].

This water problem is less critical in static HS-GC than in the purge-and-trap method. The reason for this is the much higher volumes and thus, amounts of water the analyst faces in the latter. A brief calculation can explain this situation. In the case of aqueous samples analyzed for traces of volatile halogenated hydrocarbons by purge-and-trap techniques, typical conditions are a 10–12 minute purge time at 60 °C and a flow rate of 40 mL/min. Thus, the total volume of purge gas is 400–480 mL and it will be saturated with water, representing a total water amount of 52–62 mg (saturated water vapor at 60 °C contains 0.130 g/L water, cf. Table 2-1). On the other hand, typical sample volumes in static HS-GC are in the range of 50–500 μL. If the transferred sample volume is 1 mL (i.e., 60 s with a 1 mL/min carrier gas flow), the corresponding volume of ice is 0.13 mL, and in a 0.32 mm I.D. column, this would give an ice plug only 1.6 mm long. However, ice formation and column plugging do not occur instantaneously. Our experience showed that the sampling time from an aqueous sample at 80 °C onto a 0.32 mm I.D. open-tubular column in the cryotrap can be increased up to 1.5 minutes before ice plugging may occur, and if the column used in the trap has an inner diameter of 0.53 mm, sample transfer time may even be extended to several minutes. This somewhat surprising conclusion was recognized a decade ago by Wylie [22] with his MHI technique, alluded to in Section 3.8.2. It should be clearly stated that compared to the usual sample transfer times of a few seconds (in general < 3 s), the increase of up to 100 seconds already provides a remarkable sensitivity enhancement, which is generally sufficient for most practical applications. Therefore, the problem with ice plugging is less important in the case of static HS-GC, and the examples shown in Figure 3-23 and later (Figures 8-3, 8-4, and 8-7) were carried out with no provisions against the water vapors in these samples.

However, if the sample transfer time is increased to several minutes, it becomes necessary to remove the water from the headspace gas before it enters the open-tubular column in the cryotrap. This may also be necessary with smaller sample volumes if the water causes chromatographic problems (e.g., baseline distortion) with certain detectors such as electron-capture [27], photo-ionization [27], mass spectrometric, or FTIR.

A number of methods have been developed for the removal of water from headspace samples in purge-and-trap analysis, which, however, are less suitable in static HS-GC. We describe next a system that was developed specially for static HS-GC in which water is removed by a trap containing anhydrous LiCl, coated on a porous support (e.g., 65% LiCl on Chromosorb W AW 60/80 mesh) [27]. While lithium chloride is a mild inert drying agent with a high capacity for water, it releases water already at moderate temperatures (> 120 °C): if saturated with water, it becomes a liquid and recoats the porous support, which is important for its quick regeneration. This material is packed in a small (65 mm × 0.8 mm I.D.) glass-lined tube [27].

This water trap does not completely remove the water vapor: a few percent of the saturated water vapor will also pass through. This is not a breakthrough of an excess of water, however; apparently it corresponds to the partial vapor pressure of water above LiCl: its amount was found to be constant and independent of the water load of the water trap. This small water concentration does not represent a problem.

Figure 3-24 shows this arrangement adapted to the balanced-pressure sampling system. It is an extension and a modification of the standard cryofocusing system (Figure 3-22). Sampling begins again with pressurizing the vial (I): the carrier gas flows through valves V_4 and V_1 through the sampling needle into the vial. Sample transfer is achieved when valve V_1 shuts off the carrier gas (II). Sampling is finished when the system is switched to the standby position (III) in which valve V_4 turns the carrier gas toward the T-piece

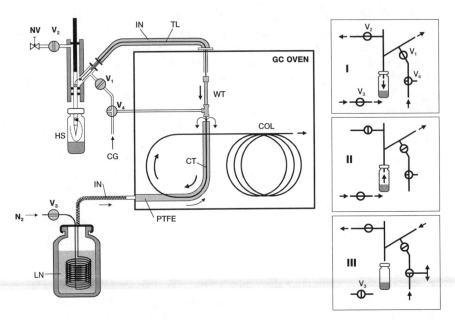

Figure 3-24. Schematic of the balanced pressure sampling system with cryofocusing and water removal by a water trap [27]. HS = headspace sampling position (see Figure 3-22), V = solenoid valves, NV = needle valve, TL = heated transfer line containing a fused silica capillary, $PTFE$ = Teflon® tube, WT = water trap, LN = liquid nitrogen bath, COL = analytical open-tubular column, CT = cryotrap, first coil of the analytical column or a separate short trap column. Valves V_1 and V_4 operate the carrier gas flow, valve V_2 operates the purge gas flow, and V_3 the cooling gas (nitrogen) flow. Position of solenoid valves: I = pressurization of the vial, II = sample transfer, III = standby (analysis).

between the water trap and the trap column in the cryotrap. In this way, the carrier gas flow to the trap column commences and, when the temperature of the column oven is increased the trapped compounds will be carried into the separation column for analysis. Simultaneously, the water trap is backflushed via the transfer line, through valve V_2. This purge gas flow can be adjusted by a needle valve (NV) at the outlet of valve V_2. The cooling gas is switched on by valve V_3 a few minutes before sample introduction to assure that the column in the trap has cooled sufficiently.

Cooling continues during sample transfer. When this is finished and the sampling needle is already in the standby position, one may continue cooling for a short time, maintaining the flow of the carrier gas through the transfer line and the water trap by having valves V_1 and V_4 open (not shown in Figure 3-24). The purpose of this step is to purge the transfer line and the water trap of any residual sample vapors and to focus them in the cryotrap. Subsequently, the coolant gas is closed by closing valve V_3 and the system switched into the standby position (III). The whole system is controlled by a microprocessor and works automatically with good reproducibility. For example, a series of 40 vials containing BTEX at the 25 ppb concentration level in water showed a relative standard deviation between 2.0 and 2.4% in an automated overnight run.

The water trap was originally developed for the headspace analysis of nonpolar volatile aromatic and halogenated hydrocarbons from aqueous samples. The chromatogram in Figure 3-25 shows as an example the determination of volatile halogenated hydrocarbons with an electrolytic conductivity detector (ELCD). Since this detector is highly halogen-specific but not very sensitive, the technique of cryofocusing is required for the determination of low ppb concentrations in water. When an electron-capture detector is used, however, this type of enrichment technique is in general not necessary for such low concentrations, as shown earlier (Figures 2-10 and 3-5), although combined with cryofocusing, ultratrace analysis can be achieved. Such an example is shown in Figure 3-26, where the concentrations of volatile halogenated hydrocarbons were detected in natural mineral water at low ppt concentrations. If we consider, for example, the peak of carbon tetrachloride (peak 3), which corresponds to a concentration of 1 ppt, we can calculate a detection limit in the (ppq) parts per quadrillion concentration range. For practical routine analysis, however, such a high enrichment can hardly be utilized due to the problems with blanks, as discussed in Section 3.3.2. Tables 3-7 and 3-8 identify the peaks in Figures 3-25 and 3-26, respectively.

Table 3-7 Identification of the peaks in Figure 3-25

Peak no.	Compound	Concentration in the sample (μg/L; ppb)
1	Trichlorofluoromethane	5.0
2	1,1-Dichloroethylene	4.3
3	1,1,2-Trilfluorotrichloroethane	5.2
4	Dichloromethane	4.5
5	trans-1,2-Dichloroethylene	4.2
6	1,1-Dichloroethane	3.9
7	Chloroform	5.0
8	1,1,1-Trichloroethane	4.2
9	1,2-Dichloroethane	4.2
10	Carbon tetrachloride	5.3
11	Trichloroethylene	4.9
12	Bromodichloromethane	6.6
13	1-Bromo-2-chloroethane	5.6
14	1,3-Dichloropropylene	4.3
15	1,1,2-Trichloroethane	4.8
16	2-Bromo-1-chloropropane	5.1
17	1,3-Dichloropropane	3.9
18	Dibromochloromethane	8.2
19	Tetrachloroethylene	5.4
20	Bromoform	9.6

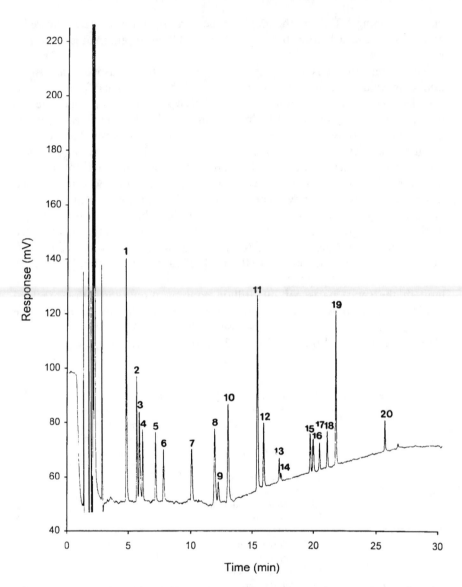

Figure 3-25. Analysis of volatile halogenated hydrocarbons in water by cryofocusing with an electrolytic conductivity detector (ELCD). *HS conditions*: Sample: 10 mL, equilibrated at 80 °C for 60 minutes, with shaker. Sample transfer time: 2 minutes, cryogenic focusing with water trap at the column head. *GC conditions*: Column: 50 m × 0.32 mm I.D. fused silica, open-tubular, coated with bonded phenyl (5%) methyl silicone stationary phase; film thickness: 2 μm. Column temperature: 10 minutes isothermal at 40 °C, then programmed at 5 °C/min to 120 °C after a 5 minute isothermal period at 120 °C, programmed at 25 °C/min to 180 °C. Carrier gas: Helium, 240 kPa. Splitless sampling. Electrolytic conductivity detector (IMT Ing. Büro, Moosbach, Germany). For peak identification and analyte concentration, see Table 3-7.

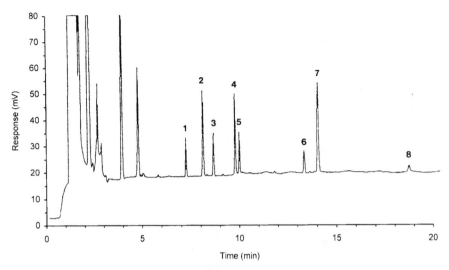

Figure 3-26. HS-GC analysis of volatile halogenated hydrocarbons in mineral water by cryofocusing with an electron-capture detector, using a water trap. *HS conditions*: 5 mL mineral water, thermostatted at 80 °C for 30 minutes with shaker. Headspace transfer time: 2 minutes, splitless. *GC conditions*: Column: 50 m × 0.32 mm I.D. fused-silica, open-tubular, coated with bonded phenyl (5%) methyl silicone stationary phase; film thickness: 2 μm. Column temperature: isothermal at 40 °C for 5 minutes, then programmed with 8 °C/min to 90 °C. Carrier gas: Helium, 160 kPa. Electron capture detector (ECD). For peak identification and analyte concentration, see Table 3-8.

Table 3-8   Identification of the peaks in Figure 3-26

Peak no.	Compound	Concentration in the sample (ng/L; ppt)
1	Chloroform	8
2	1,1,1-Trichloroethane	6
3	Carbon tetrachloride	1
4	Trichloroethylene	5
5	Bromodichloromethane	11
6	Dibromochloromethane	7
7	Tetrachloroethylene	30
8	Bromoform	—

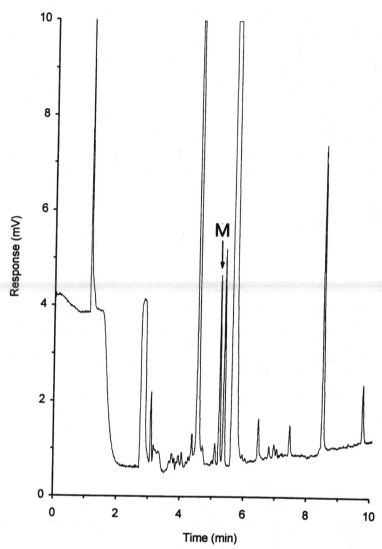

Figure 3-27. HS-GC analysis of 0.1 ppm (mg/L) methanol (*M*) in a pharmaceutical aqueous buffer solution by cryofocusing, using a water trap. *HS conditions*: 2 mL aqueous sample + 2 g K_2CO_3, thermostatted at 80 °C for 60 minutes. Headspace transfer time: 5 minutes, splitless. Cryotrap: 55 cm × 0.32 mm I.D. fused-silica, open-tubular column coated with bonded phenyl (5%) methyl silicone stationary phase, film thickness: 1 µm. *GC conditions*: Column: 50 m × 0.32 mm I.D. fused-silica, open-tubular, coated with bonded poly(ethylene glycol) stationary phase; film thickness 0.4 µm. Column temperature: isothermal at 40 °C for 1 minute, then programmed at 20 °C/min to 55 °C and after 2 minutes at 6 °C/min to 180 °C. Carrier gas: Helium, 175 kPa, 3.4 mL/min. Flame-ionization detector.

According to our experience [27], even polar volatile analytes up to boiling points of about 130 °C pass through the water trap into the cryogenic trap. An example is shown in Figure 3-27, where 0.1 ppm methanol was determined quantitatively in an aqueous buffer solution for pharmaceutical applications. Reactive compounds, such as volatile free fatty acids, however, did not pass the water trap and were completely retained. The water trap is at the temperature of the GC oven and, since it is not a chromatographic system, low volatile compounds may be adsorbed. However, if the water trap is purged further for a short time (e.g. 1 min) after the end of sample injection, even such adsorbed compounds may be swept downstream and subsequently trapped again in the cryo-trap from which all trapped compounds are finally eluted together.

We have mentioned the general problem in gas chromatography of finding a compromise between sensitivity and resolution. In HS-GC this compromise is undoubtedly the technique of cryofocusing. As an illustration of the effect of the water trap in combination with cryogenic trapping, Figure 3-28 shows the

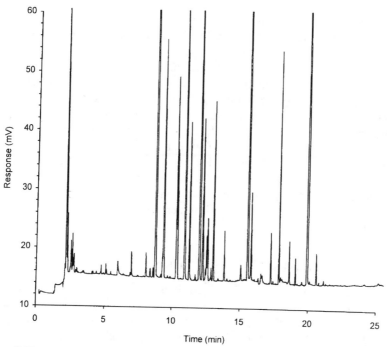

Figure 3-28.   HS-GC analysis of fennel seed, with cryofocusing, using a water trap. *HS conditions*: Sample: 250 mg fennel seed, at 25 °C. Headspace transfer time: 9.9 minutes, splitless sampling. Cryo-trap: 120 cm × 0.25 mm I.D. open-tubular column coated with bonded cyanopropyl (14%) methyl silicone (OV-1701) stationary phase; film thickness: 1 μm. *GC conditions*: Column: 15 m × 0.18 mm I.D. fused-silica, open-tubular, coated with OV-1701 stationary phase; film thickness: 1 μm. Column temperature: isothermal at 40 °C for 4 minutes, then programmed at 5 °C/min to 90 °C and then at 8 °C/min to 120 °C. Carrier gas: Helium, 79 kPa. Flame-ionization detector.

headspace analysis of fennel seed, with a sample transfer time of as long as 9.9 minutes using a narrow-bore, open-tubular column with an inner diameter of 0.18 mm. In spite of this long sampling time, the peaks are very sharp, and highly efficient separation was achieved. Indeed, the possibility of increasing the sensitivity by applying cryofocusing now makes it possible to keep the sample at room temperature thus to avoid excessive heating. Any thermal stress during adsorption and desorption as in the case of dynamic headspace techniques is strictly avoided here. This is particularly important for aroma analysis, where temperature-sensitive compounds often decompose and the resulting artifacts may be mistaken for true sample components.

3.9 Reaction Headspace–Gas Chromatography

A headspace vial can conveniently be used as a reactor to carry out chemical reactions and to monitor the resulting volatile products. Reactions are often carried out to generate volatile derivatives from polar compounds by simple derivatization reactions. Another possibility is to generate volatile compounds from nonvolatile parent molecules by degradation or by liberation from a biological conjugate.

Most chemical reactions are equilibrium systems; therefore, it is advantageous that the reaction products of interest usually have a higher volatility and thus are removed from the liquid reaction medium by vaporization into the headspace, shifting the chemical reaction to completeness.

We now outline various possibilities of reaction HS-GC by summarizing some selected applications each illustrating a certain type of reaction that may be carried out in the headspace vial. However, it is outside the scope of this book to describe the reaction conditions in detail. The discussion is confined to reactions that are performed *in* the vial and on-line with headspace sampling. External derivatization techniques carried out separately are beyond our scope.

3.9.1 Derivatization in the Headspace Vial

The formation of more volatile derivatives is a general technique widely used in gas chromatography, and it can also be applied in HS-GC, using the headspace vial as the reaction vessel. Such derivatization reactions as esterification, transesterification, acetylation, silylation, and alkylation are simple chemical reactions; however, they may present some problems when used in headspace analysis. Since, in general, the derivatization reagents are volatile compounds also and are added in excess amounts, they may interfere in the chromatograms; also a derivatization reaction is not applicable in HS-GC if other volatile reaction products are produced in excessive amounts. This set of properties can represent a problem (e.g., in the preparation of ethyl esters by adding an excess of ethanol) as a result of the generation of excessive pressure in the vial. The reagent may also contain a number of impurities or may generate peaks in the chromatogram by reaction with other, originally non-volatile compounds in the matrix. Finally, many derivatization reactions require water-free conditions, which restricts the possibilities with some practical samples.

As mentioned, essentially any derivatization reaction resulting in volatile products can be used in HS-GC, and Sections 3.9.1.1 through 3.9.1.4 give only a brief summary of the most frequently used derivatives. For details on derivatization in general and the reactions used in GC, the reader is referred to the specialized books dealing with this subject [28,29].

3.9.1.1 Methylation

A very universal procedure for methylation involves the use of dimethyl sulfate, in the presence of a base and/or potassium carbonate [30]. Dimethyl sulfate reacts in aqueous solution at a high pH value with many polar groups, including alcohols, diols, phenols, acids, and amines. For example:

$$R\text{---}OH + (CH_3O)_2SO_2 \xrightarrow{\text{K}_2\text{CO}_3 + \text{KOH}} R\text{---}OCH_3$$

The reaction is fast, while the excess of the reagent is slowly decomposed by inorganic ions such as chloride, cyanide, sulfide, and carbonate, and finally also by water. This permits the selection of conditions under which the excess of the reagent disappears and does not disturb the chromatogram.

Good recoveries of about 90% have been reported for phenols, and even pentachlorophenol can be analyzed in water solution at a concentration of 50 ng/L with an electron-capture detector. The recoveries for carboxylic acids increase with increasing chain length, but the method fails for dicarboxylic acids. A structure dependence was also found with diols: the 1,2-diols (1,2-ethanediol and 1,2-propanediol) do not react satisfactorily, but with increasing distance of the hydroxy groups the reaction yield becomes quantitative (e.g., for 1,4-butanediol). Example 3.4 presents the conditions for the methylation of diols in aqueous solution.

Example 3.4

First 6 g of K_2CO_3 is placed in the headspace vial, and two pellets of KOH are added, followed by 5 mL of the aqueous sample solution and 0.1 mL of dimethyl sulfate. The vial is immediately crimp-capped and equilibrated for one hour at 80 °C with shaking. Although the methylation of the diols is reasonably fast, the one-hour thermostatting is necessary to destroy the excess reagent.

An important warning is needed here: *Dimethyl sulfate is carcinogenic and extremely hazardous!* It has a very high toxicity, and inhalation of its vapors can cause pulmonary edema. It is, therefore, worthwhile considering less toxic alternatives. A number of other reagents have been described for methylation [28], but their use in HS-GC is little known.

3.9.1.2 Esterification

Esterification of free fatty acids to prepare the more volatile and less polar esters is a general reaction used in gas chromatography:

$$RCOOH + R'OH \xrightarrow{\text{catalyst}} RCOOR' + H_2O$$

The reaction needs a catalyst such as an acid (sulfuric or hydrochloric acid), methyl iodide, boron trichloride, or boron trifluoride [31], of which particularly the last was found to be preferable. A convenient alternative to sulfuric acid is solid $NaHSO_4$, which is easier to handle and — a particularly useful property in HS-GC — has an additional salting-out effect [32]. The reaction can be carried out in aqueous solution or by the total vaporization technique (see Section 4.6.1), where finally all the compounds, except the catalyst, are in the vapor phase. An example is shown in Figure 3-29 illustrating the determi-

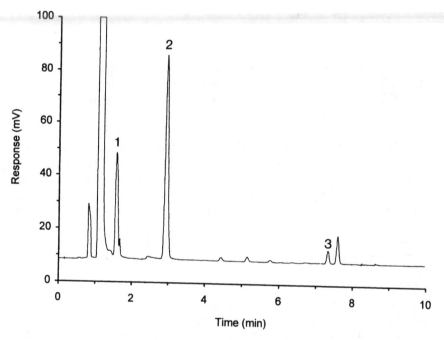

Figure 3-29. Determination of organic acids from sugar molasses by esterification. *HS conditions*: Sample: 500 µL aqueous solution (10%) of sugar molasses + 100 µL saturated $NaHSO_4$ solution + 50 µL ethanol, thermostatted at 80 °C for 1.5 hours with shaking. *GC conditions*: Column: 25 m × 0.32 mm I.D. fused-silica, open-tubular, coated with bonded methyl silicone stationary phase; film thickness: 5 µm. Column temperature: isothermal at 60 °C for 5 minutes, then programmed with 8 °C/min to 150 °C, then with 20 °C/min to 250 °C. Carrier gas: Helium, 135/100 kPa, split sampling: 0.08 minute. Flame-ionization detector. *Peaks*: Ethyl esters of 1 = formic acid (0.23 mg/mL), 2 = acetic acid (0.61 mg/mL), 3 = n-butanoic acid (0.026 mg/mL).

nation some volatile free fatty acids from sugar molasses as the corresponding ethyl esters, prepared from aqueous solution at 80 °C. A quantitative analysis has shown that the reaction yield of the ethyl ester formation was close to 100%, due to the higher volatility (lower partition coefficients) of the resulting ethyl esters. The reaction is not very fast, however: a long equilibration time of 1.5 hours was found to be necessary. Although these acids are present at high concentrations, they could not be analyzed directly without esterification unless enriched cryofocusing, due to the low volatility from aqueous solution. The preparation of the more volatile esters here has the additional advantage that formic acid, which is not detected by a flame-ionization detector if present as the free acid, can also be determined.

The total vaporization technique (TVT) is preferable if the resulting esters are not very volatile and therefore, need a high temperature ($>100\,°C$): if a two-phase system with an aqueous solution were present in the vial, it would generate excessive internal pressures at such a temperature. In the TVT, the sample volume should not exceed $15\,\mu L$ (for a $22.3\,mL$ vial), assuring a single-phase system. An example given in Chapter 8 (see Figure 8-9) shows the analysis of dicarboxylic acids from *Bacteroides fragilis*.

3.9.1.3 Transesterification

The transesterification reaction, used particularly to prepare the volatile fatty acid (methyl) esters from the glycerides present in fats and lipids, is as follows:

$$R\text{—COO-glyceride} + CH_3OH \xrightarrow{\ CH_3ONa\ } R\text{—COOCH}_3 + \text{glycerol}$$

Again, a catalyst is needed for the reaction, and usually sodium methoxide in methanolic solution is used for this purpose.

A good example for the utilization of this reaction is the HS-GC determination of the amount of butterfat in milk chocolate [33]. The triglycerides of butterfat contain butanoic acid, and the transesterification yielding the volatile butanoic methylester can be readily carried out in the headspace vial: Figure 3-30 shows the resulting chromatogram.

Example 3.5

To 100 mg of the chocolate sample in the headspace vial, 2 mL of dry methanol containing 0.5% sodium methoxide is added, and the mixture is thermostatted at 70 °C for 1.5 hours.

Naturally, a number of the higher fatty acids homologous to butanoic acid are also present in the sample, and their methyl esters will also be formed. However, they are removed from the chromatogram by backflushing.

3.9.1.4 Acetylation

Another simple reaction is acetylation with acetic anhydride or similar anhydrides. A good example is the determination of glycerol in a polywax sample.

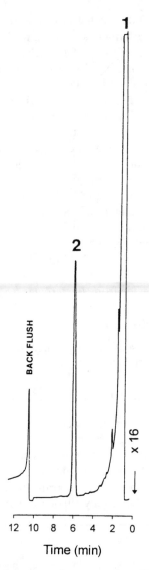

Time (min)

Figure 3-30. Measurement of the butterfat content in milk chocolate, by transesterification. *HS conditions*: Sample: 100 mg chocolate + 2 mL dry methanol containing 0.5% CH$_3$ONa, thermostatted at 70 °C for 1.5 hours. *GC conditions*: Two 1 m × 1/8 in. O.D. packed columns containing 0.1% SP-1000 on Carbopak C 80/100 mesh. Column temperature: 140 °C, isothermal. Column backflushed at 10 minutes. Flame-ionization detector. *Peaks*: 1 = methanol, 2 = butanoic acid methyl ester (0.71%).

Extraction with water could not be applied here because of the good water solubility of the polywax matrix. Therefore, glycerol was converted to triacetin:

$$C_3H_5(OH)_3 \xrightarrow{\text{(CH}_3\text{CO)}_2\text{O}} C_3H_5(OOC\text{---}CH_3)_3$$

Triacetin boils at 259 °C, and it has sufficient vapor pressure at 120 °C to be measured by HS-GC. Figure 3-31 shows the obtained chromatogram. Sodium

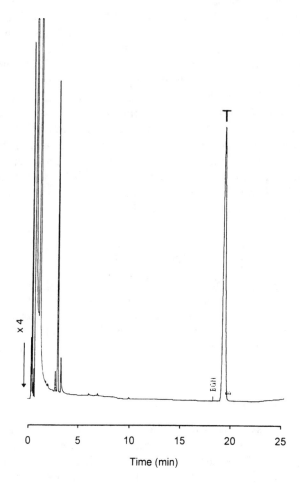

Figure 3-31. Determination of glycerol in a polywax sample, by acetylation to triacetin. *HS conditions*: Sample: 1 mg polywax + 100 mg dry Na_2CO_3 + 100 μL acetic anhydride, thermostatted at 120 °C for 90 min. *GC conditions*: Column: 25 m × 0.32 mm I.D. fused-silica, open-tubular, coated with OV-1701 cyanopropyl (7%) phenyl (7%) methyl silicone stationary phase; film thickness: 1 μm. Column temperature: 2 minutes isothermal at 80 °C, then programmed at 8 °C/min to 120 °C, and 20 °C/min to 200 °C. Carrier gas: Helium, 158/130 kPa, split sampling. Sample transfer time: 1.2 seconds. Flame-ionization detector. *Peak*: T = triacetin (9.8%).

carbonate (water free) was added to bond the excess acetic acid released during the reaction.

3.9.2 Subtraction HS-GC

In the preceding examples an essentially nonvolatile or lowvolatile analyte was modified to obtain a highly volatile derivative that could be analyzed by HS-GC. However, the opposite is also possible: to render some volatile sample components nonvolatile. This "subtraction HS-GC" may be useful if the peaks of these compounds would obscure the peaks of the analytes of interest in the chromatogram.

An example is shown in Figure 3-32, where free fatty acids, alcohols, and acetates in beer were analyzed (chromatogram A). Due to the low volatility and low concentration in beer, the technique of cryofocusing was necessary. After the addition of NaOH (chromatogram B) the peaks of the free acids disappeared, as did the esters (e.g., 2-phenylethyl acetate), while the peaks of the alcohols (e.g., 2-phenylethanol, furfuryl alcohol) remained unchanged. The small peak in chromatogram B under the peak of acetic acid (peak 1 in chromatogram A) is a nonidentified compound (probably an alcohol) and not acetic acid; this is known because a blank run with acetic acid in alkaline solution showed no peak at all, thus excluding also any memory peak of acetic acid.

$$R{-}COOR' \xrightarrow{\text{NaOH}} R{-}COONa + R'OH$$

The esters are saponified, forming also the nonvolatile sodium salts, while the formed alcohols are volatile and can be determined by HS-GC.

There are many such possibilities, as shown by a comprehensive review on such subtraction methods in gas chromatography by Berezkin [34]. Although not specifically described for HS-GC, many of these reactions may also be carried out in the headspace vial.

3.9.3 Special Reactions

The preceding discussion dealt with generally applicable chemical reactions. In addition, however, a number of special reactions also described in the literature were utilized to modify certain specific compounds or compound groups. As an exception to our general philosophy in this book, we are listing here specific applications that illustrate the various fields in which the use of HS-GC may be advantageous. For the specific conditions, the referenced publications should be consulted.

It was mentioned earlier (Section 3.9.1.1) that the methylation of *vicinal diols* by dimethyl sulfate is not very satisfactory. On the other hand, the determination of glycols in water and soil at low ppm concentrations is of some interest, particularly to determine soil pollution at airports from deicing of airplanes,

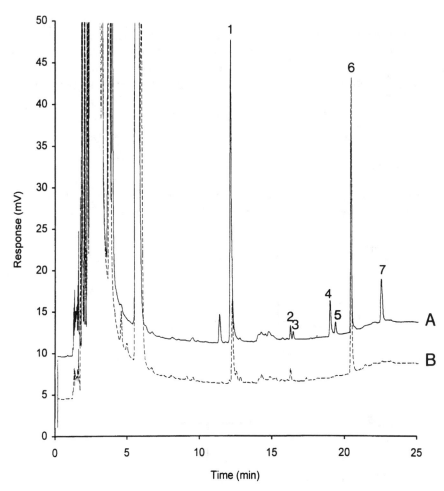

Figure 3-32. Identification of esters and volatile free acids in beer with cryofocusing by subtraction-HS-GC. *HS conditions*: Sample: 0.5 mL beer + 100 μL water (A) or NaOH solution (B), equilibrated at 60 °C for 60 minutes. Sample transfer time: 2 minutes. Cryo-trap: 1 m × 0.32 mm I.D. polar deactivated fused-silica guard capillary (Restek). *GC conditions*: Column: 50 m × 0.32 mm I.D. fused-silica, open-tubular, coated with SP-1000 stationary phase; film thickness: 1 μm. Column temperature: initial temperature: 60 °C, then programmed at 25 °C/min to 80 °C, then isothermal for 6 minutes, then programmed at 8 °C/min to 200 °C. Carrier gas: Helium, 215 kPa, splitless sampling. Flame-ionization detector. Figure A: *Peaks*: 1 = acetic acid, 2 = furfuryl alcohol, 3 = isovaleric acid, 4 = 2-phenylethyl acetate, 5 = hexanoic acid, 6 = 2-phenylethanol, 7 = n-octanoic acid. Figure B: free acids and esters disappeared after the addition of NaOH solution to the beer sample.

an operation in which glycols are widely used. Since the headspace analysis of ethylene glycol is accompanied by wall adsorption problems (see Section 3.1.4) and is not sufficiently sensitive (see Figure 3-3), the use of derivatives looks promising. Reaction with phenylboronic acid produces volatile cyclic compounds [35], and this procedure was adapted here for headspace analysis:

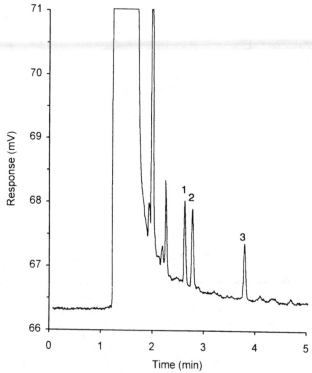

Figure 3-33 shows the determination of 10 ppm glycol in water by TVT using 2 μL of the aqueous sample together with 2 μL of the reagent.

Figure 3-33. Determination of glycols in water after derivatization and total vaporization. *HS conditions*: Sample: 2 μL aqueous sample + 2 μL reagent solution (0.41 mmol phenylboronic acid in 2,2-dimethoxypropane), vaporized at 120 °C for 15 minutes. *GC conditions*: Column: 50 m × 0.53 mm I.D. fused-silica, open-tubular, coated with bonded methyl silicone stationary phase; film thickness: 1 μm. Column temperature: 8 minutes isothermal at 165 °C, then programmed at 20 °C/min to 250 °C. Carrier gas: Helium, 155/127 kPa, split sampling. Sample transfer time: 0.04 minute. Flame-ionization detector. *Peaks*: Derivatives of (10 ppm each) 1 = ethylene glycol, 2 = 1,2-propanediol, 3 = 1,3-propanediol (internal standard).

Metaldehyde was determined in animal tissues by treating the sample with 6 M sulfuric acid: metaldehyde depolymerized into the volatile acetaldehyde [36]:

$$(CH_3CHO)_n \xrightarrow{\text{H}_2\text{SO}_4} CH_3CHO$$

The determination of *carbon monoxide* in blood bonded to hemoglobin (CO-Hb) is of interest in the case of fire victims due to smoke inhalation, and an official German analytical procedure [37] utilizes reaction HS-GC for its determination: CO is liberated in the closed headspace vial by reaction with potassium hexacyanoferrate(III) (potassium ferricyanide) and determined with a flame-ionization detector, after conversion to methane in a catalytic reactor (see Section 5.8):

$$CO\text{—}Hb \xrightarrow{\text{K}_3\text{Fe(CN)}_6} CO \xrightarrow{\text{H}_2\text{(catalyst)}} CH_4$$

Figure 3-34 shows the determination of 17 ppm carbon monoxide in the blood of a cigarette smoker.*

Another blood analysis method utilizing reaction HS-GC is the determination of cyanide. The cyanide bonded in blood is converted to free hydrogen cyanide by the addition of glacial acetic acid, which is then determined using a Porapak Q column with a thermionic (NPD) detector [38]:

$$R\text{—}CN \xrightarrow{\text{(CH}_3\text{CO)}_2\text{O}} HCN$$

In a similar way, cyanides and thiocyanides present in water can also be determined, in the concentration range of 0.01–100 ppm [39].

Pesticides are in general not volatile enough to be analyzed by HS-GC. Dithiocarbamates (Maneb, Zineb, Mancozeb, Mezined, Ferbam, Manam, Propineb, sodium dimethyl dithiocarbamate, Thiram), which are widely used as fungicides in flower, fruit, and vegetable cultivation, are nonvolatile and therefore not amenable for use in even normal GC analysis. Upon degradation with stannous chloride/hydrochloric acid solution, however, the highly volatile carbon disulfide is formed as a common derivative from all dithiocarbamates. For example, in the case of Maneb:

$$
\begin{array}{c}
\overset{\displaystyle S}{\overset{\displaystyle \|}{CH_2\text{—}NH\text{—}C}} \\
\Big| \qquad\qquad\quad \diagdown \\
\qquad\qquad\qquad\quad Mn \xrightarrow{\text{SnCl}_2 + \text{HCl}} CS_2 \\
\Big| \qquad\qquad\quad \diagup \\
CH_2\text{—}NH\text{—}\underset{\displaystyle \underset{\displaystyle S}{\|}}{C}
\end{array}
$$

The formed carbon disulfide can be detected either with a flame-photometric detector (FPD) [40,41] or an electron-capture detector [42], with the

---

* For this analysis the official method [37] was modified by using two short Carbosieve SII columns in backflush configuration for faster GC analysis and the MHE external standard method for calibration.

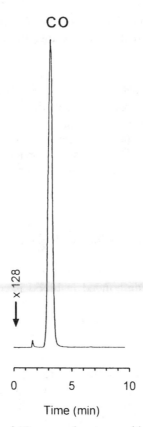

Figure 3-34 Determination of 17 ppm carbon monoxide (CO) in the blood (CO-Hb) of a cigarette smoker, transforming the liberated CO into methane, using a methanizer. *HS conditions*: Sample: 1 mL blood + 1 mL aqueous (30%) $K_3Fe(CN)_6$ solution, thermostatted at 50 °C for 30 minutes. *GC conditions*: Column: Two 0.5 m × 1/8 in. O.D. packed, containing Carbosieve SII 60/80 mesh;. Column temperature: 40 °C, isothermal. Column backflushed at 10 minutes. Flame-ionization detector.

respective detection limits of about 25 and 10 μg/kg. The chromatogram in Figure 3-35 shows the determination of 2 μg/g Maneb in the leaves (100 mg) of a carnation. The reaction can be carried out with a few grams of sample (e.g., food) together with 5 mL of the reagent solution (1.5% stannous chloride in 5N aqueous hydrochloric acid) in the headspace vial, with equilibration for one hour at 70 °C [40] or 80 °C [41], while shaking. The only restriction is that HCl vapors may result in severe corrosion problems: therefore, the use of an inert and metal-free headspace sampling system is mandatory. If this is not available, a two-step procedure may be applied [42]. First the reaction is carried out in a closed vial, at a higher temperature, outside the HS instrument; then, upon completion of the reaction, the vial is transferred to the headspace sampler and equilibrated at 40 °C, at which temperature the corrosion problems are much less severe.

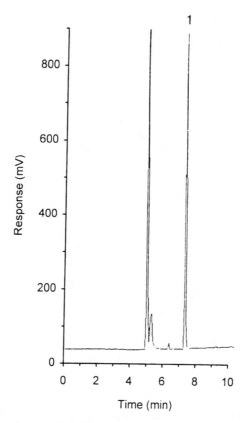

Figure 3-35. Determination of dithiocarbamate (Maneb) in carnation leaves by reduction to carbon disulfide. *HS conditions*: Sample: Two carnation leaves + 5 mL reagent solution (1.5% $SnCl_2$ in 5 M HCl), 1 hour at 100 °C min for the reaction, then thermostatted at 50 °C for 30 minutes. *GC conditions*: Column: 50 m × 0.53 mm I.D. fused-silica, open-tubular, coated with bonded phenyl (5%) methyl silicone stationary phase; film thickness: 1 μm. Column temperature: 50 °C isothermal. Sample transfer time 0.08 minute. Split sampling. Electron capture detector with nitrogen make up gas. *Peak*: 1 = carbon disulfide.

Enzymatic reactions can also be carried out in the headspace vial. For example, β-galactosidase activity was determined with 2,2,2-trichloro-ethyl-β-D-galactopyranoside as the substrate, followed by headspace analysis of the liberated 2,2,2-trichloroethanol with detection by an electron capture detector. This assay was then applied to the quantitative determination of *Escherichia coli* bacteria [43]. The determination of 3-hydroxybutyrate in plasma was carried out by reaction HS-GC, after conversion to acetone by the use of 3-hydroxybutyrate dehydrogenase/lactate dehydrogenase, coupled with acetoacetate decarboxylate [44]:

$$CH_3\!-\!\underset{\underset{\displaystyle OH}{|}}{CH}\!-\!CH_2\!-\!COOR \xrightarrow{\text{enzyme}} CH_3\!-\!\underset{\underset{\displaystyle O}{\|}}{C}\!-\!CH_3$$

Another enzymatic reaction was described to determine the benzene intake of the human body by measuring the phenol content of urine. Phenol is the metabolite of benzene, and in urine it is present as either phenyl sulfate or phenyl glucoronide. Using β-glucoronidase/sulfatase as the hydrolyzing enzyme, free phenol can be liberated and determined by HS-GC [45]. The hydrolysis procedure was carried out in the closed-headspace vial containing the sample (1 mL), an acetate buffer (0.2 mL), and the enzyme preparation (50 μL) at 40 °C for 16 hours (overnight). After this time the closed vial was transferred into the HS-GC instrument and analyzed without any further treatment. This technique was later modified by replacing the slow enzymatic hydrolysis by hydrolysis with sulfuric acid [46]: in this case a one-hour thermostatting period at 75–80 °C was enough to free the phenols. The technique was used for the routine determination of the ratio of phenol to p-cresol in urine, which changes significantly when the intestinal flora in humans is disturbed.

Even *metal–organic compounds* are amenable to HS-GC analysis. For example, organomercury compounds were determined in biological samples, after transformation into methylmercury iodides by iodoacetic acid [47]:

$$R\!-\!Hg\!-\!CH_3 \xrightarrow{\text{ICH}_2\text{COOH}} CH_3\!-\!HgI$$

An inert, metal-free sample path was found to be essential for the headspace sampling in this application [48].

Organic arsenic compounds were determined by HS-GC for toxicological purposes, after reduction to arsenic hydride with sodium borohydride, using a thermionic (NPD) detector [49]. Arsenobetain ([Btn]As) was determined in fish and seafood by the reaction with NaOH, forming trimethylarsine oxide. This was finally reduced by sodium borohydride to trimethylarsine, which then could be determined by HS-GC, with a flame-ionization detector [50,51]:

$$[Btn]As \xrightarrow{\text{NaOH}} (CH_3)_3AsO \xrightarrow{\text{NaBH}_4} (CH_3)_3As$$

3.9.4 HS-GC Analysis of Volatile Derivatives from Inorganic Compounds

Naturally, the most important volatile inorganic compound, present in practically every sample, is water. It may be determined directly by HS-GC, using a thermal-conductivity detector; this is an attractive alternative to the Karl–Fischer titration for both liquid and solid samples [52] (cf. Example 5.6). However, water can also be determined through certain reaction products: for example, after reaction with 2,2-dimethoxypropane, by measuring the formed acetone, using a flame-ionization detector [53]:

$$H_2O + CH_3\!-\!\overset{\displaystyle OCH_3}{\underset{\displaystyle OCH_3}{\overset{|}{\underset{|}{C}}}}\!-\!CH_3 \rightarrow CH_3\!-\!\overset{\displaystyle O}{\overset{\|}{C}}\!-\!CH_3 + 2CH_3OH$$

The well-known reaction of water with calcium carbide forming acetylene

$$H_2O + CaC_2 \rightarrow C_2H_2 + CaO$$

can also be used, determining the amount of acetylene by HS-GC [54].

It is emphasized that all methods for determining the amount of water in a sample must reflect awareness that the determination is primarily influenced by the water blank due to the inherent humidity of the air enclosed in the vial (cf. Table 2-1 and Example 5.6), and the ratio of the area (height) of the water peak from the sample to that of the blank will fundamentally limit the minimum detectable quantity (MDQ). For this reason there is no particular advantage to using the higher detector sensitivity of an FID for organic water derivatives.

Inorganic anions can also be analyzed by reaction HS-GC. In addition to polar organic compounds, dimethyl sulfate reacts with several anions, forming the corresponding methyl derivatives (e.g., acetonitrile from cyanide) [30]:

$$R\!-\!CN \xrightarrow{\ (CH_3O)_2SO_2\ } CH_3\!-\!CN$$

Ionic fluoride is of particular interest for studies of caries and fluorosis. The standard method for its determination [55] is based on silylation: ionic fluoride reacts with trimethylchlorosilane in acidic medium, forming the highly volatile and water-resistant trimethylfluorosilane:

$$F^- + (CH_3)_3SiCl \xrightarrow{\ H^+\ } (CH_3)_3SiF + Cl^-$$

In the standard method the reaction product is extracted with a solvent; however, the determination can also be carried out by direct headspace analysis. This method has been used in the HS-GC determination of fluoride in cosmetics and pharmaceuticals [56] and in fluorinated milk [57], by the headspace analysis of the formed silane. Figure 3-36 shows the chromatogram obtained in the determination of fluoride in tea with this technique. This reaction might also be used for the determination of hydrogen fluoride in the atmosphere, by first absorbing an air sample into an alkaline solution.

Figure 3-36. Determination of inorganic fluoride in a tea sample. *HS conditions*: Sample: 100 mg tea suspended in 1 mL water + 1 mL concentrated HCl + 5 µL trimethylchlorosilane, thermostatted at 80 °C for 30 minutes. *GC conditions*: Column: 50 m × 0.32 mm I.D. fused-silica, open-tubular, coated with bonded phenyl (5%) methyl silicone stationary phase; film thickness: 5 µm. Column temperature: 5 minutes isothermal at 60 °C, then programmed at 20 °C/min to 120 °C. Carrier gas: Helium, 117 kPa. Splitless sampling. Sample transfer time: 2.4 seconds. Flame-ionization detector. *Peak*: F = trimethylfluorosilane (formed from the inorganic fluoride in the sample).

References

1. B. Kolb, D. Boege, P. Pospisil, and H. Riegger, German Patent 28-34-186 (1980).
2. *Chromatography Catalog No. 350*, Alltech Associates, Inc., Deerfield, IL, 1995.
3. C. Sadowski, and J. E. Purcell, *Chromatogr. Newslt.* **9**, 52–55 (1981).
4. P. J. Gilliver and H. E. Nursten, *Chem. Ind. (London)* **1972**, 541.
5. *ASTM D-4526–85* (91): Standard Practice for Determination of Volatiles in Polymers by Headspace Gas Chromatography.
6. N. R. Litvinov, T. M. Lyutova, and Yu. G. Sukhareva (1976), described in B. V. Ioffe, and A. G. Vitenberg, *Head-Space Analysis, and Related Methods in Gas Chromatography*, Wiley, New York, 1982; p. 70.
7. *HSS-3A/2B* Headspace Analysis Systems, Shimadzu Corp., Kyoto, Japan.
8. Z. Penton, *HRC*, **17**, 647–650 (1994).
9. D. Jentzsch, H. Krüger, G. Lebrecht, G. Dencks, and J. Gut, *Fresenius Z. Anal. Chem.* **236**, 96–118 (1968).
10. W. Closta, H. Klemm, P. Pospisil, R. Riegger, G. Siess, and B. Kolb, *Chromatogr. Newslt.* **11**, 13–17 (1983).
11. H. Pauschmann, *Chromatographia*, **5**, 622–623 (1972).
12. G. Göke, *Chromatographia*, **3**, 376–377 (1970).
13. Courtesy of T. A. Berger (Hewlett–Packard Co., Wilmington, DE), and P. A. Kester (Tekmar Co., Cincinnati, OH).
14. M. Kuck, presentation at the Second International Colloquium on Gas Chromatographic Headspace Analysis, Überlingen, October 18–20, 1978, in B. Kolb (editor), *Applied Headspace Gas Chromatography*, Heyden & Son, London, 1980; pp. 12–22.
15. B. Kolb, P. Pospisil, T. Borath, and M. Auer, *HRC/CC*, **2**, 283–287 (1979).
16. L. S. Ettre, J. E. Purcell, J. Widomski, B. Kolb, and P. Pospisil, presentation at EXPOCHEM 1979, Houston, TX, October 22–25, 1979; *J. Chromatogr. Sci.* **18**, 116–125 (1980).
17. J. V. Hinshaw, and W. A. Seferovic, presentation at the 17th International Symposium on Capillary Chromatography, and Electrophoresis, Wintergreen, VA, May 7–11, 1995; *Proceedings*, pp. 188–189.
18. B. Kolb, P. Pospisil, and M. Auer, German Patent DE 31-17173-C3 (1989).
19. *U. S. Pharmacopeia XXIII:* Organic Volatile Impurities (467). Method IV, 1995; pp. 1746–1747.
20. B. Kolb, B. Liebhardt, and L. S. Ettre, *Chromatographia*, **21**, 305–311 (1986).
21. J. P. Pankow and M. E. Rosen, *HRC/CC*, **7**, 504–508 (1984).
22. P. L. Wylie, *Chromatographia*, **21**, 251–258 (1986).
23. G. Takeoka, and W. Jennings, *J. Chromatogr. Sci.* **22**, 177–184 (1984).
24. K. Grob Jr., *J. Chromatogr.* **213**, 3–14 (1981).
25. K. Grob, and A. Habich, *HRC/CC*, **6**, 34–35 (1983).
26. W. F. Burns, D. T. Tingley, and R. C. Evans, *HRC/CC*, **5**, 504–505 (1982).
27. B. Kolb, G. Zwick, and M. Auer, *HRC*, **19**, 37–42 (1996).
28. K. Blau and G. S. King (editors), *Handbook of Derivatives for Chromatography*, Heyden & Son, London, 1977.
29. J. Drozd, *Chemical Derivatization in Gas Chromatography*, Elsevier, Amsterdam, 1981.
30. H. J. Neu, W. Zimer, and W. Merz, *Fresenius J. Anal. Chem.* **340**, 65–70 (1991).
31. A. Akan, S. Fukushima, K. Matsurbara, S. Takahashi, and H. Shiono, *J. Chromatogr.* **529**, 155–160 (1990).
32. S. Heitefuss, A. Heine, and S. H. Seifert, *J. Chromatogr.* **532**, 374–378 (1990).
33. B. Kolb, D. Matthes, and M. Auer, *Applications of Gas Chromatographic Headspace Analysis No. 25*, Bodenseewerk Perkin-Elmer, Überlingen, 1979.
34. V. G. Berezkin, *Fresenius Z. Anal. Chem.* **296**, 1–17 (1979).
35. W. H. Porter, and A. Auansakul, *Clin. Chem.* **28**, 75–78 (1982).
36. C. J. Griffiths, *J. Chromatogr.* **295**, 240–247 (1984).

37. D. Henschler (editor), *Analysen in biologischen Material*, Vol. 2: Analytische Methoden zur Prüfung gesundheitsschädlicher Arbeitsstoffe, 11th ed., Verlag Chemie, Weinheim, 1994.

38. J. Zmaecnik, and J. Tam, *J. Anal. Toxicol.* **11**, 47–48 (1987).

39. G. Nota, V. R. Miraglia, C. Improta, and A. Acampora, *J. Chromatogr.* **207**, 47–54 (1981).

40. T. K. McGhie, and P. T. Holland, *Analyst*, **112**, 1075–1076 (1987).

41. Report by the Panel on the Determination of Dithiocarbamate Residues, Ministry of Agriculture, Fisheries, and Food (United Kingdom), *Analyst*, **106**, 782–787 (1981).

42. M. J. M. Jongen, J. C. Ravensberger, R. Engel, and L. H. Leenheers, *J. Chromatogr. Sci.* **29**, 292–296 (1991).

43. B. Koppen, and L. Dalgaard, *J. Chromatogr. Sci.* **321**, 385–391 (1995).

44. M. Kimura, K. Kobayashi, A. Matsuoka, K. Hayashi, and Y. Kimura, *Clin. Chem.* **31**, 596–598 (1995).

45. E. R. Adlard, C. B. Milne, and P. E. Tindle, *Chromatographia*, **14**, 507–509 (1981).

46. W. Tashkov, L. Benchev, N. Rizov, M. Kafedzhieva, and A. Kolarska, *Chromatographia*, **32**, 466–468 (1991).

47. P. Lansens, C. C. Laino, C. Meuleman, and W. Baeyens, *J. Chromatogr.* **586**, 329–340 (1991).

48. P. Lansens, and W. Baeyens, *HRC/CC*, **12**, 132–133 (1989).

49. W. Vycudilik, *Arch. Toxicol.* **36**, 177–180 (1976).

50. U. Ballin, Ph. D. Thesis, University of Hannover, Department of Chemistry (1992).

51. U. Ballin, R. Kruse, and H. A. Rüssel-Sinn, *Arch. Lebensmittelhyg.* **42**, 27–32 (1992).

52. B. Kolb, and M. Auer, *Fresenius Z. Anal. Chem.* **336**, 291–296, 297–302 (1992).

53. G. Scharfenberger, Colloquium über die gaschromatographische Dampfraumanalyse, October 20–21, 1975, Überlingen.

54. J. M. Loeper, and R. L. Grob Jr., *J. Chromatogr.* **463**, 365–374 (1989).

55. J. Fresen, F. Cox, and M. Witter, *Pharm. Weekbl.* **103**, 909 (1968).

56. J. Hild, Colloquium über die gaschromatographische Dampfraumanalyse, October 3–4, 1983, Bad Nauheim.

57. W. Tashkov, I. Benchev, N. Rizov, and A. Kolarska, *Chromatographia*, **29**, 544–546 (1990).

CHAPTER

4

Sample Handling in HS-GC

In headspace analysis the sample usually (except in the specially prepared single-compound standards) consists of a mixture of volatile and nonvolatile components. We are interested in either the qualitative composition of the volatile part of the sample or the amount (concentration) of some or all the volatile components present. Usually we have a two-phase (gas–liquid or gas–solid) system in the sample vial, and we analyze an aliquot of the gas phase (the headspace). The advantage of this situation is that as far as the gas chromatography is concerned, we do not have to worry about the nonvolatile sample components: they will not enter the gas chromatograph. This is a particular advantage of HS-GC over general GC.

The samples analyzed are usually liquids or solids. Sometimes we dissolve the solid and analyze the solution. We may also analyze gas samples, either by collecting a gas sample in the vial or by fully evaporating a small sample volume in the sample vial.

While by HS-GC we can directly analyze the volatile compounds present in a complex sample, the *matrix* of the sample needs particular consideration. Let us not forget that the quantitative analysis depends on the distribution of the analyte(s) between the two phases and, naturally, the composition of the sample phase has an important influence on the partition coefficient of the analyte. This question is particularly important in the preparation and analysis of *standards*. Just as in general gas chromatography, one needs a calibration (response) factor to establish the corresponding amount (concentration) from the peak area. However, contrary to general GC, this factor does not depend solely on the detector's response to the particular compound, hence cannot be obtained simply by analyzing a known amount of the analyte: the *calibration factor* will also depend on the matrix. Therefore, in general, the sample matrix must be reproduced if quantitation is accomplished by comparison with the analysis of a known amount (concentration) of the analyte. Alternatively, there are methods that may be used to reduce or eliminate the matrix effect.

Due to the different nature and volatility of the analytes, the manner of introduction of a sample into the headspace vial is very important. Similarly, because of the importance of the standards used in quantitative analysis, their preparation needs special consideration.

As discussed in Chapter 2, quantitative reproducibility and comparison of individual analyses (e.g., sample vs. a standard) is possible only if an equilibrium* is established for the volatile analyte in the sample vial between the two phases prior to transferring an aliquot into the gas chromatograph. Thus, we must first investigate the questions related to equilibration.

4.1 Equilibration

Essentially the time needed for equilibration depends on the diffusion of the volatile sample components from and into the sample proper. One can indicate some general guidelines concerning the equilibration time; however, it cannot be predicted. Therefore, in the case of an unknown sample, particularly if no previous information is available, the time needed must be established. This can be done by preparing a number of vials with the same sample, thermostatting them for different times but otherwise using identical conditions, and plotting the obtained peak area against the thermostatting time. An example is given in Figure 4-1: the time needed for equilibration is the time until a constant peak area is obtained. Particularly useful for such time-consuming kinetic measurements is the automated *progressive mode* of thermostatting. For details see Section 3.4.2. The equilibration time is the shortest time the sample has to be thermostatted. The analytical result will not change if the actual thermostatting time is longer than the time needed for equilibration; excessively long thermostatting should be avoided, however, because some samples may be sensitive to prolonged heating.

The equilibration time may be fairly long—in fact, generally longer than the analysis time. To start the equilibration of the second sample only after the analysis of the first has been accomplished, will result in an unnecessary loss of time and sample throughput. To eliminate this unnecessary time delay for fast routine analysis, modern automated instruments can thermostat each sample for the same time and start it while the preceding sample is still being analyzed. This is the *overlapping constant mode* of thermostatting (see Section 3.4.2 and Figure 3-8).

---

* As will be discussed in Chapter 7, in special cases one may work under nonequilibrium conditions.

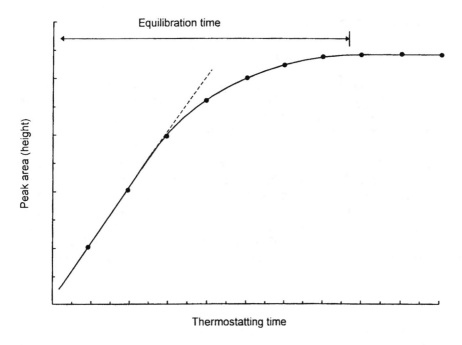

Figure 4-1. Thermostatting time versus peak area (height) by *progressive working mode.*

4.1.1 Gas Samples

HS-GC can also be used to analyze gas samples. These may have two origins: either they were collected as a gas (see Section 4.3.1), or a very small volume of a liquid sample was completely evaporated in the vial (TVT: see Section 4.6.1).

Although one would consider a gas sample to be already at equilibrium, it is still important to maintain a certain equilibration time for two reasons. First, the sample must be kept at a constant temperature during the whole analytical procedure; and second, condensation from the original gaseous sample may occur if the original temperature of the collected sample was higher than the temperature to which it was exposed during transport and storage, and such a condensed sample again needs some time for total evaporation.

In the case of complete evaporation of the sample in the vial (total vaporization technique: see Section 4.6.1), it is important to realize that evaporation is not instantaneous. Therefore, a certain "equilibration time" is always necessary. Since complete evaporation is usually applied for the preparation of standards, the most convenient way is to carry it through the same equilibration process as the original sample.

4.1.2 Liquid Samples

4.1.2.1 *General Properties*

The shortest equilibration times are found with liquid samples of low viscosity (e.g., aqueous solutions), but the equilibration time always depends on the sample volume. This is illustrated in Figure 4-2, which plots the peak area obtained from the HS analysis of toluene in an aqueous solution as a function of the sample volume. The remarkable increase in the equilibration time from 2 mL up to 10 mL of sample can be explained by the increasing diffusion path of the analyte in the matrix. Thus, from this point of view, one would prefer small sample volumes. On the other hand, the need to use larger sample volumes can be explained by referring to the fundamental relationship given earlier between headspace sensitivity (peak area, A), the original sample concentration C_o, the partition coefficient K of the analyte, and the sample volume (phase ratio β: the ratio of the headspace volume to sample volume in

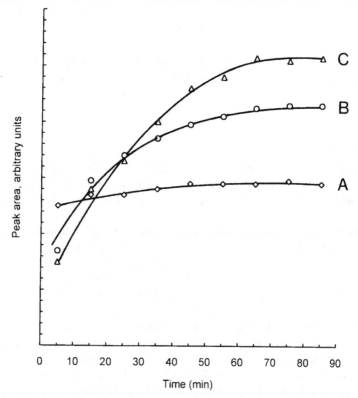

Figure 4-2. Equilibration time of aqueous solutions of toluene (2 ppm), with varying volume. Thermostatting temperature: 60 °C. Headspace vial volume: 22.3 mL. Sample volume: $A = 2.0\,mL$, $B = 5.0\,$mL, $C = 10.0\,$mL.

the vial):

$$A \propto C_G = \frac{C_o}{K + \beta} \tag{2.19}$$

Analytes having a low solubility in the matrix will have a small partition coefficient (e.g., $K = 1.86$ for toluene in water at 60 °C [1]). In such a case, the phase ratio (i.e., the sample size) has a strong influence on the peak area obtained for a given original sample concentration. Thus, it is advantageous to have a small phase ratio (i.e., use a large sample volume). It can be calculated* for this particular case that increasing the sample volume from 2 mL ($\beta = 10.15$) to 10 mL ($\beta = 1.23$) will decrease the denominator of eq. (2.19) and thus, increase the peak area obtained, by a factor of 4. Therefore, in such cases, the use of sample volumes corresponding to 50% of the vial's volume is not unusual.

In the case of analytes having a high solubility, hence a higher partition coefficient, the influence of the sample volume on headspace sensitivity is negligible. Considering aqueous solutions of ethanol at 60 °C ($K = 511$ [2]), the respective values of the denominator of eq. (2.19) for sample volumes of 2 and 10 mL are 521.5 and 512.2: this means that a fivefold increase in sample volume would increase the peak area only by 1.8%, which is negligible. Thus, for such samples there is no reason to use larger sample volumes.

4.1.2.2 Reduction of the Equilibration Time for Liquid Samples

Equilibration of liquid samples can be speeded up by raising the equilibration temperature. Einstein's law states that the diffusion coefficient† D is proportional to the absolute temperature T:

$$D \approx T/f \tag{4.1}$$

where f is the so-called friction factor, the value of which is proportional to the size of the analyte's molecule. However, since this effect is proportional to the absolute temperature, it is hardly possible to take advantage for practical application: an increase of the sample temperature from 60 °C to 90 °C reduces the equilibration time by only about 8%.

A better way to decrease the time needed for equilibration is *continuous mixing* of the sample in the vial during the equilibration process. This can be done, for example, by shaking, and most commercial HS-GC systems provide this function.When applying a shaker, it is important that the sample be in resonance with the shaker frequency to obtain the desired mechanical mixing

---

* In this and subsequent calculations, the vial's volume is always taken as 22.3 mL.
† The diffusion coefficient expresses the mass of the analyte diffusing through a plane of 1 cm² in one second.

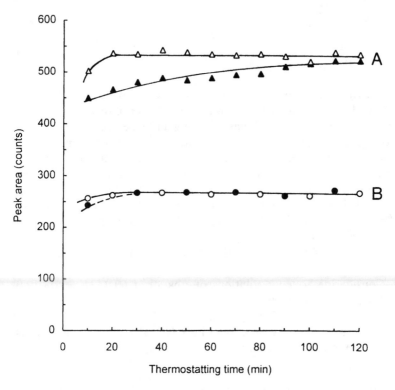

Figure 4-3. Equilibration times of aqueous solutions of toluene (*A*) and acetone (*B*) without (solid symbols) and with (open symbols) shaking. Thermostatting temperature: 60 °C. Sample volume: 5.0 mL of *A* (8 ppm solution of toluene) and *B* (150 ppm solution of acetone).

effect. However, in the case of liquid samples, the resonance frequency depends on the sample viscosity and its volume. To overcome this problem, the shaker's frequency should be automatically varied during equilibration within a broad range, so that each sample in the thermostat gets in resonance.

We recommend using a shaker in the determination of nonpolar volatile organic compounds (VOCs) in aqueous solutions exceeding a volume of about 3 mL. Figure 4-3 illustrates the effect of agitation on the equilibration time for toluene and acetone in an aqueous solution of 5 mL in a 22.3 mL vial: without shaking, the state of equilibrium for toluene is approached very slowly and needs about 2 hours. On the other hand, shakers have not much influence on polar compounds, as can be seen in the curves for acetone. This difference may be due to the higher partition coefficient and the corresponding smaller concentration in the gas phase relative to the concentration in the sample. The difference of the peak areas between using the shaker or not becomes smaller too and may already be hidden in the bandwidth of the precision.

4.1.3 Solid Samples

In the case of solid samples, diffusion takes much longer. This can be seen by comparing the order of magnitude of the diffusion coefficient D: its order of magnitude is 10^{-6} in liquids and 10^{-8} to 10^{-11} in solids, while it is 10^{-1} in gases. Therefore, long equilibration times are often characteristic of solid samples, where the diameter and thickness of solid particles will determine the time of diffusion. However, the porosity of the solid sample is also important, and apparently, its surface also influences the speed of equilibration. Therefore, some porous solid samples with a high surface area often have surprisingly short equilibration times.

Solid samples can be classified into two groups [3]. In the first case, under certain conditions the solid behaves as a *partition system*: in other words, distribution of the volatile analyte is governed by its partition coefficient, which is constant over a wide concentration range. In general, polymers that can be heated above their respective glass transition temperatures behave in this way. Such samples may be analyzed as solids (*solid approach*) or dissolved in an organic solvent and analyzed as a solution (*solution approach*, discussed in Section 4.2).

According to the definition in ASTM E 1142 [4], "glass transition" (second-order transition) refers to the reversible change of a amorphous polymer structure from a hard and relatively brittle condition to a viscous and rubbery condition (or vice versa) with an increase of the speed of diffusion. Generally this transition occurs over a relatively narrow temperature region. The glass transition temperature is usually specified as the approximate midpoint of the temperature range over which the glass transition takes place. Table 4-1 lists glass transition temperatures for a number of polymers [5,6].

If the solid approach is used, it is always important to ascertain that a partition system is present. This can be seen, for example, by a linear MHE plot (see Section 5.5). Therefore it is advisable in the case of a new sample to first test the possibility of its analysis by multiple headspace extraction, even if at the end a less time-consuming quantitative analytical method is selected for routine measurements. This is demonstrated in Chapter 6, which gives examples for method development.

As mentioned, equilibration of a solid sample in general, but not necessarily always, takes longer than the equilibration of a liquid sample, and it depends on the structure of the solid and also on the equilibration temperature. Figure 4-4 compares the equilibration of residual styrene monomer present in various polystyrene samples, illustrating also the effect of temperature. In the case of a polystyrene foam, the porous structure favors quick vaporization if the temperature is sufficiently high, and such a situation is achieved by thermostatting at 120 °C. At lower temperatures, even in the case of the porous foam, at least 2.5 hours is needed to achieve equilibrium. On the other hand, in the case of polystyrene granules or pellets, no equilibrium could be reached at 120 °C: the 4 mm × 3 mm polystyrene pellets have too long a diffusion path for

Table 4-1   Glass transition temperatures T_g of some polymers

Polymer	T_g (°C)	Note	Ref.
Silicone rubber	−125		6
Dimethyl silicone	−123		5
cis Polybutadiene	−99		6
cis 1,4-Polyisoprene	−83		6
Natural rubber	−63		6
Poly(butyl acrylate)	−55	atactic	5
Acrylonitrile–butadiene copolymer	−41	random	6
Styrene–butadiene copolymer	−39	random	6
Neoprene	−38		6
Polypropylene	−6	isotactic	6
Poly(vinyl acetate)	+29	atactic	5
Nylon-6	+51	fiber	6
Nylon-66	+59	fiber	6
Poly(vinyl chloride)	+81		5
Polyester	+85	fiber	6
Polystyrene	+100	isotactic and atactic	5
Polyacrylonitrile	+101	fiber	6
Poly(methyl methacrylate)	+105	atactic	5
Polycarbonate	+149		5
Polyamide–polyimide	+279		6

monomeric styrene to be liberated from the polymer matrix. The same polystyrene sample, present as a powder after freeze-grinding, has an acceptable equilibration time of about 2 hours at a thermostatting temperature of 120 °C. About the same time is required for an aqueous polymer dispersion (emulsion) at 90 °C. Such a dispersion is a three-phase system, and the time-limiting process is the slow diffusion from the polymer droplets into the surrounding aqueous matrix, from which the further partitioning process into the gas phase then proceeds much faster. In any case it should be noted that such polymer emulsions no longer need the classical "solution approach" [7], with its accompanying reduced sensitivity.

To decrease the diffusion path, the solid sample should be divided into small pieces (e.g., by cutting). Freeze-grinding under cooling with solid carbon dioxide or liquid nitrogen is often required to avoid loss of the volatile compounds due to heat generation during the milling process. This should be confirmed during method development by comparison with an independent method (e.g., the solution approach: see below), if such a method is available. However, there is often no alternative at all.

In the preceding example we saw that raising the thermostatting temperature can reduce the time needed for equilibration. However, in the case of polymers, one must be careful to avoid beginning of depolymerization or side reactions: for example, if residual ethylene oxide is to be determined in sterilized clinical PVC material, these polymer samples should not be heated

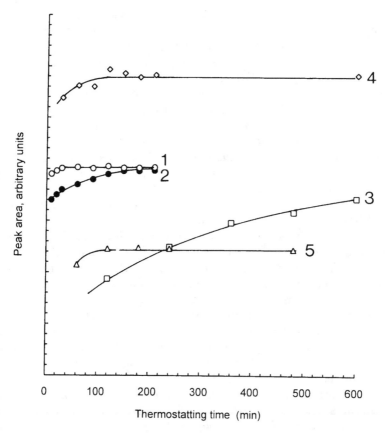

Figure 4-4. Equilibration time/temperature of residual monomeric styrene in various polystyrene samples. *1* = Polystyrene foam thermostatted at 120 °C; *2* = polystyrene foam thermostatted at 100 °C; *3* = polystyrene granules (4 mm × 3 mm) thermostatted at 120 °C; *4* = polystyrene powder obtained from polystyrene granules by freeze-grinding thermostatted at 120 °C; *5* = aqueous polymer dispersion, thermostatted at 90 °C.

above 100–120 °C to avoid splitting off HCl and the formation of chlorohydrin.

Starting with the pioneering work of Romano on the determination of residual ethylene oxide used for sterilization [8], the solid approach has been used in a wide range of applications. The determination of residual vinyl chloride monomer in poly(vinyl chloride) resins is another classical example for this approach [9,10].

The other form of solids occurring in HS-GC analysis represents *adsorption systems*: in fact, most solids, particularly those with a high surface area, show some adsorption effects. Whenever these are involved, phase distribution of the volatile analytes is determined and controlled by the adsorption coefficient, which is not independent of the concentration but is a function of it. Therefore, adsorption systems have only a very narrow linear concentration range if any at all. The limited or nonlinear phase distribution is caused by inhomogeneous two-dimensional distribution of adsorption sites with different adsorption energies across the sample surface.

Another peculiar characteristic of adsorption systems is the remarkable difference in equilibration time. The time needed until a steady-state equilibrium is achieved is rather long, much longer than with comparable liquid samples or solid samples representing a partition system. This is due to the slow diffusion processes in the micropores of the sample. In addition, the time function of equilibration also depends on concentration: lower concentrations of the analyte are adsorbed more strongly (hence longer) on the strong adsorption sites of the samples. This phenomenon explains the particular shape of the MHE plot discussed in Section 5.5.6, under case F.

In general one can state that a sample representing a nonlinear adsorption system cannot be quantitatively analyzed in its original form, due to its nonlinear behavior. The necessary modifications involve altering the surface properties of the material. This is possible by, for example, the addition of a small amount of a modifier: by blocking the strong adsorption sites, the modifier transfers the sample into a weak adsorption system with homogeneous adsorptivity and an extended linear concentration range. When the amount of the liquid displacer (modifier) is increased, the system begins to change into a partition system; further increase in the amount of the added liquid displacer results in the buildup of a separate liquid phase in which the analyte is dissolved. In this way, the solid sample is transformed into a liquid sample, with a well-defined matrix (the added solvent); and the remaining solid particles, now suspended in the liquid, have no further influence on the partitioning between the two phases (gas and liquid)

Questions associated with the quantitative headspace analysis of adsorption-type solid samples are be discussed in Section 5.6.

4.2 Solution Approach

If the solid sample is soluble in an organic solvent or in water, the determination can be simplified using the *solution approach*: it is dissolved in the solvent and the solution is analyzed by HS-GC. The equilibration time will depend on the viscosity of the resulting solution and will not necessarily be shorter than it would be if the solid sample were analyzed directly. For example, vinyl chloride monomer (VCM) equilibrates from a porous PVC resin in less than 30 minutes if heated above the glass transition point of 85 °C, while from a viscous solution in an organic solvent the equilibration time usually exceeds one hour. Naturally one would try to achieve a high concentration of the dissolved polymer to get the necessary high sensitivity for the volatile monomer in it, but the payoff is the combination of high viscosity with long thermostatting time. If, however, the solid polymer is present in the form of pellets or granules, with accompanying excessive equilibration times, the traditional solution approach is undoubtedly an advantageous alternative to the technique of freeze-grinding, provided the reduced sensitivity is acceptable. The earliest example for the solution approach was described by Rohrschneider [11], who determined residual styrene monomer (SM) in polystyrene (PS) granules by this technique, dissolving the granules in dimethylformamide (DMF) (about 200 mg PS in 2 mL DMF) and using n-butylbenzene as the internal standard. In this way, 1 ppm SM could be determined in PS. Figure 4-5 (based on the data given by Rohrschneider) compares the equilibration times of PS granules versus the DMF solution, at 75 °C. As seen, the granules would need more than 20 hours for equilibration, whereas the solution required only about 100 minutes.

While the solution approach is a convenient method for sample preparation, it has some disadvantages. The headspace sensitivity will be reduced compared to the direct analysis ("solid approach") of the solid sample. Since the volume

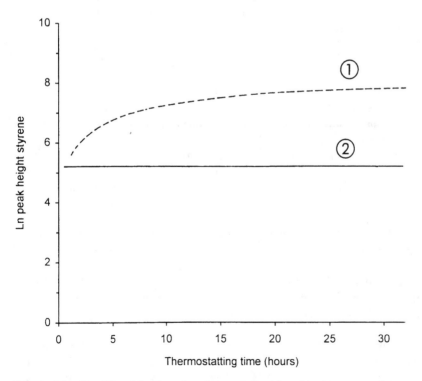

Figure 4-5. Equilibration time for the analysis of residual monomeric styrene in polystyrene (PS) samples. Thermostatting temperature: 75 °C. *1* = PS granules; *2* = DMF solution.

Source: reproduced with permission of L. Rohrschneider and *Zeitschrift für analytische Chemie* [11].

of the solvent (solution) will be at least one order of magnitude higher than that of the solid sample, the concentration of the analyte in the resulting solution will be less than in the original (solid) sample.

Another negative effect of the solution approach is related to the nature of the organic solvents used, which generally have a high solubility for the analyte. In such solvents the partition coefficient — which is usually relatively small in a solid sample matrix — will be large; on the other hand, it is clear from eq. 2.19 that an increase in K will further reduce C_G hence the headspace sensitivity.

Let us consider as a practical example the determination of 100 ppm of a solvent in a solid drug sample in which the original partition coefficient K should be 10. If we use the "solution approach" for the determination of OVIs (see Figure 3-19) according to the USP [12] by preparing a solution of one gram in 5 mL, we dilute the sample and at the same time the partition coefficient will increase, assuming as an example an increase from 10 up to 100. Therefore, we have to consider two effects, the dilution effect and the decrease in the volatility. We can compare the influence on the headspace sensitivity by applying eq. 2.19, where C_o is the analyte's concentration in the sample, K is the partition coefficient, and β is the phase ratio (ratio of headspace volume to sample volume in the vial).

$$C_G = \frac{C_o}{K + \beta} \tag{2.19}$$

One gram of the original solid sample, assuming a density of 1.0, if placed in a 22.3 mL vial, gives a phase ratio of $\beta = 21.3$. The same sample, dissolved in 5 mL of solvent, assuming now a partition coefficient of $K = 100$, has a phase ratio of $\beta = 3.46$ and the original concentration in the solid sample (C_o^S) of 100 ppm is decreased to 20 ppm in the resulting solution (C_o^L). If we compare the influence on the resulting concentration in the gas phase C_G by using eq. 2.19, we can calculate the ratio of the gas phase concentrations of the solid sample C_G^S to that of the liquid solution C_G^L and obtain $C_G^S/C_G^L = 16.5$. The solid approach therefore is 16.5 times more sensitive. The only reason for analyzing a solid sample by the "solution approach" is the advantage conferred by the better presuppositions for quantitative analysis of a liquid compared to a solid sample. These aspects are discussed in Chapter 5.

The reduced sensitivity due to the increase of the partition coefficient in the organic solvent may be improved by a modification of the solution approach first described by Steichen [13], who used it in the analysis of residual monomers such as 2-ethylhexyl acrylate (EHA) in polymers. We have already mentioned this possibility in Section 2.3.4: after dissolving the polymer sample in a water-miscible solvent (e.g., DMA), water is added to the solution. In this way the partition coefficient will be reduced, but the analyte will still remain in solution. For example, Steichen reported a 600-fold increase in the sensitivity for EHA obtained in this way, by adding 3 mL of water to 4 mL of the DMA solution.

With respect to the solution approach, it is imperative that the solvent used be of the highest possible purity. Even very low impurity concentrations may interfere with the chromatographic separation.* Therefore, the solvent must be carefully investigated for possible interference prior to use. This is particularly important if the sensitivity will need to be improved further by applying the cryofocusing technique.

Most of the officially recommended headspace procedures utilize the solution approach. For example, it is used in some ASTM methods [14,15] and recommended by the U.S. Pharmacopeia [12] in the analysis of organic volatile impurities in various drugs. Still, whenever possible, the direct analysis of the solid samples is to be preferred.

4.3 Sample Handling and Sample Introduction

Here we deal with the methods by which samples are prepared and transferred into the headspace vial for analysis.

4.3.1 Gas Samples

As already mentioned, headspace–gas chromatography may also be used for the analysis of gaseous (vapor) samples when a condensed phase is absent: the vial now contains only a single phase. In essence, this is no longer "headspace" analysis and the vial only serves as a container for the sample.

Gas (vapor) samples may have two origins: either they were collected as such, sampling an atmosphere or some other gas (automobile exhaust, breath, etc.) or a very small amount of a (liquid) sample was fully evaporated in the vial.

The collection of an air sample is particularly straightforward: in fact, if an open vial is kept for some time at the sampling place, it will be filled with the ambient atmosphere by diffusion. The collection time can be shortened by withdrawing the original air present in the vial by means of a small hand pump (Figure 4-6). After a few pump strokes, the air in the vial is exchanged with the outside atmosphere. If the vial is closed in the usual way with a septum and a crimped cap, the sample remains stable until analysis [16]. This "vial sampling technique" is comparable to the air sampling procedure using large passivated stainless steel canisters [17]. The vial can be considered to be a minicanister; in fact, it has the advantage of being a disposable item, and thus one does not encounter the problems of contamination and sample carryover known with the stainless steel canisters, which must be cleaned after each use.

---

* For example, one of the reasons Romano [8] selected the solid approach for the determination of residual ethylene oxide in polymer tubes was that acetone, which would be used in the solution approach, always had an impurity, with a retention time very close to that of ethylene oxide.

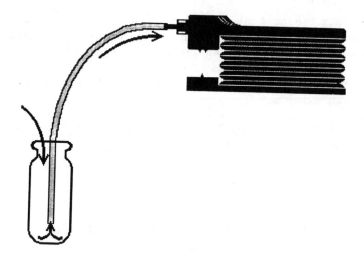

Figure 4-6. Collection of ambient air sample in a headspace vial [16]

Another way to exchange the air in the vial with a gaseous sample is to deliver the latter from a tube, which is inserted into the open vial just as a liquid sample can be introduced with a syringe using the open-vial technique. A special application of this technique is the analysis of breath: the breath sample is blown into the vial through a plastic tube inserted into the vial, similar to the use of a syringe needle for transfer of a liquid sample by the open vial-technique (see Figure 3-7). After the vial has been filled with the gas sample, the tube is rapidly withdrawn and the prepared seal pressed on the vial and crimp-capped. The reproducibility may depend somewhat on the skill of the operator, but the risk of sample losses during the whole procedure is less than it is with alternative techniques (e.g., the septum is punctured twice with two sampling needles, one for the inlet and one for the outlet of the gas sample) [18].

Whenever the container is flushed with gas, it works as an exponential gas dilution device, and the process can be described by a first-order equation:

$$C_e = C_o \cdot e^{(-Ft/V)} \qquad\qquad (4.2)$$

where C_o is the original and C_e is the actual concentration of the original gas (air) filling the container at time t; V is the volume of the container, and F is the flow of the new gas into the container. Thus, we can calculate how long it would take to exchange the atmosphere of a vial:

$$t = -\frac{V}{F} \cdot \ln(C_e/C_o) \qquad\qquad (4.3)$$

For example, if 95% of the air in the vial is to be replaced, then $C_e/C_o = 5/100 = 0.05$, while in the case of 99% replacement, $C_e/C_o = 0.01$. Assuming

$F = 50\,\text{mL/min}$ and $V_V = 2.3\,\text{mL}$:

95% replacement: $t = 1.2$ min

99% replacement: $t = 1.8$ min

Quantitative determination with a gas sample can be carried out by means of an external standard (see Section 5.3). The easiest method for preparation of the external standard is the use of the total vaporization technique (see Section 4.6.1). It was mentioned in Section 4.1.1 that even a gas sample needs a certain equilibration time.

4.3.2 Liquid Samples

Liquid samples must be handled with particular care, considering the high volatility of some analytes in a particular matrix. For this reason, the possibility of using pipettes for sample transfer is limited, because they are particularly prone to introduce errors due to losses by vaporization. Here we have to distinguish between analytes with low and high partition coefficients.

The use of pipettes for sample transfer is not recommended in the case of highly volatile analytes with low partition coefficients because of the possibility of sample loss due to vaporization.* For example, if the partition coefficient of aromatic hydrocarbons (BTEX: benzene, toluene, ethylbenzene, and the xylenes) is less than 5, there is always a risk of a falsified analysis when a pipette is used. Therefore, such samples should be handled and transferred with syringes or syringe-type devices, carefully avoiding the formation of any gas space in the syringe. If this rule is not followed, a nonlinear calibration function will result in the event, for example, of a stock solution further diluted by a stepwise dilution procedure. The same consideration applies for sample containers that always should be filled completely, without any remaining headspace in them. The responsible analyst should reject as unreliable bottles only half-filled with such a sample.

The situation is different with samples with a high partition coefficient. If, for example, a bottle is half-filled with blood for the analysis of ethanol, the amount of ethanol in the gas space of the bottle is about 0.1% of its amount in the blood and the use of a pipette would result in such a small further loss that the accuracy of the analysis would be affected hardly at all.

In many cases improper sample handling is the cause of poorer than expected precision. This can be checked by preparing a standard test sample containing two compounds, one with low and the other with high partition coefficient. An aqueous solution of ethanol and toluene represents a good example: their partition coefficients are (cf. Table 2-2):

ethanol: 40 °C: $K = 1355$ 60 °C: $K = 511$

toluene: 40 °C: $K = 2.82$ 60 °C: $K = 1.77$

---

* It is emphasized that this problem is not limited to headspace analysis, but is common in any quantitative measurement involving highly volatile materials.

If we carry out a number of HS-GC determinations with this sample and find good precision for ethanol ($<1\%$) but unsatisfactory ($>2\%$) results for toluene, we have a strong indication of inadequate sample handling. Only if the precision for ethanol is also poor can some instrumental problem be suspected.

The partition coefficient of most compounds in organic solvents is relatively high, and in such cases pipettes may be used to transfer aliquots from one solution to another or into the headspace vial without any problem.

Sometimes neither pipettes nor syringes can handle viscous samples. In such cases the sample should first be diluted with a suitable solvent, because otherwise it may become difficult to reproduce the sample volume (the phase ratio), and thus calibration becomes more complicated (see Section 5.7).

4.3.3 Solid Samples

Typical solid samples for headspace analysis are polymers (resins, pellets, granules), printed packaging material (plastic or aluminum films), pharmaceuticals (drugs and excipients), and soil. They can be analyzed either as solids (*solid approach*) or in solution (*solution approach*), and questions associated with these techniques were summarized earlier. Here we deal with the sample handling procedure: how the original bulk of a solid material may be processed for direct analysis or for dissolution.

Naturally, a special question associated with such samples entails the division of a bulk sample to get a representative aliquot. This is a general consideration in analytical chemistry, and standardization bodies (ASTM, DIN, etc.) have developed guidelines.

In general, solid polymer samples are too large to permit the achievement of equilibrium in any reasonable time (if it is possible at all). This follows from the Einstein equation [19], which relates the time of diffusion t to the diffusion coefficient D and the diffusion path length d:

$$t = d^2/2D \tag{4.4}$$

Therefore, if for any reason the solution approach cannot be applied, the only choice is to comminute the material mechanically, by grinding. To avoid excessive heat generation and the concomitant loss of volatiles, *freeze-grinding* is necessary, and the sample must be chilled with solid carbon dioxide or with liquid nitrogen. Such manipulation can reduce considerably the size of the material. For example, in the case of poly(ethylene terephthalate) (PET) samples, which are particularly difficult to comminute, freeze-grinding at liquid nitrogen temperature for 10 minutes resulted in 0.15–0.42 mm particles [20], while according to the F. Kurt Retsch Company grinding 5 mm PET pellets for 2 minutes at 15,000 rpm under liquid nitrogen with their Type ZM1000 Ultracentrifugal Mill results in the following particle distribution [21]:

$<315\,\mu m$	80.8%
$315–500\,\mu m$	16.6%
$>500\,\mu m$	2.6%

The possible extent of particle reduction by grinding can be evaluated with help of eq. 4.4. If d_1 is the original diffusion path and it is d_2 after grinding, then

$$t_2 = t_1(d_2/d_1)^2 \tag{4.4a}$$

For example, reduction from 4 mm to 0.315 mm results in an almost three-orders-of-magnitude reduction of the time needed for diffusion ($t_2 = 0.0014\,t_1$).

The polymer powder obtained from freeze-grinding can be placed directly into the vial for headspace analysis by the solid approach (see Section 4.1.3).

It should be noted that if a powdered polymer sample is stored in a closed bottle for a longer time, it must be mixed by shaking before an aliquot is taken out for analysis. The reason for this is that a concentration gradient may have formed during storage, since the upper layers of the powder release the volatiles faster in the closed bottle into the headspace, even if stored at room temperature.

Thin printed plastic films can be cut into small pieces and analyzed directly. However, in the case of thick, laminated films, freeze-grinding may again be necessary to reduce the equilibration time. Sample collection and selection is critical in the case of such films, and various standards [22,23] describe exact methods for the selection of the representative sample. The approximate size of the final aliquot placed into the vial depends on the vial's volume. The German standard [23] recommends the use of the following pieces for analysis:

Vial volume (mL)	Film area (cm^2)
6	15
20–25	25–50
50	50–100

while the ASTM standard [22] advises cutting an 8×36 in. (2 sq. ft = 203×915 mm) piece with help of a template and then cutting this in strips of 1 in. (25.4 mm) width before placing them in the headspace vial. Naturally, such large samples would need much larger sample vials, and the quoted ASTM standard—which describes manual operation, using 5-mL gas syringes to transfer a headspace aliquot into the gas chromatograph—utilizes one-liter flasks for this purpose. However, with up-to-date automated HS-GC instruments, such a determination can be carried out with smaller samples (see Section 6.3). These recommendations therefore must be modified and adjusted to the available equipment and, as outlined earlier, it is the phase ratio β that must be constant for equal headspace sensitivity rather than the absolute size of vials or samples. On the other hand, a bigger vial can accept a larger and thus more representative aliquot from a printed film.

4.4 Preparation of Standard Solutions

In general, standard solutions for calibration and identification should be prepared freshly by the analyst, because certified standards are rare and are not sufficiently stable, considering the high volatility of the analytes in HS-GC. For quantitative analysis such standard solutions are required at low concentrations and usually, the original stock solution needs to be further diluted by a solvent, or one may use the MHE technique.

The main reason for using a standard *solution* rather than the pure compound is that the latter is usually to be added to the sample, to a certain matrix (as an internal or external standard), or to a vial for total vaporization (see later: Section 4.6.1) in such a small amount that its handling would be difficult; on the other hand, when microsyringes or micropipettes are used, even very small volumes of a solution can be handled easily and with good accuracy. A good example is the determination of tetrachloroethylene (TCE) in olive oil found in Chapter 5 (see Figure 5-6), where a pure oil with added TCE was used as the external standard. The TCE level investigated was around 1 ppm and the sample volume was 5 mL: to prepare an external standard by direct addition of TCE to an oil would involve the accurate handling of a few nanoliters of TCE, which is practically impossible. On the other hand, one can easily prepare a diluted TCE stock solution in a suitable solvent with a concentration of e.g., 81.6 μg/mL by the 1:100 dilution of a primary standard solution of 50 μL (81.5 mg) TCE per 10 mL solvent — and then add 10 μL of this solution (containing 0.816 μg TCE) to 5 mL of pure olive oil. The resulting TCE concentration in the soil will be 0.16 ppm (wt/vol).

4.4.1 Preparation of a Standard Solution from a Liquid or Solid Substance

Considering the preparation of a stock solution, the main problem is the purity of the solvent and its interference with other peaks in the chromatogram. In general, high-boiling solvents are preferred which elute late in the chromatogram and can be backflushed in the column, an important point in routine work to reduce analysis time. The purity of the solvent must always be checked by GC analysis prior to use. Table 4-2 lists some of the solvents we have been using in our work [24,25].

The general procedure to prepare a stock solution from a liquid or solid substance involves adding a specified amount (volume) of the pure compound into a small (e.g., 10 mL) ground-glass, stoppered volumetric flask. When preparing a stock solution from a multicomponent mixture, it is wise to begin the addition with the compound of the lowest volatility and to proceed following the line of decreasing boiling points (increasing volatility). Next the flask is filled up to volume with the solvent and stoppered, and its content mixed by inverting the flask several times. This stock standard solution should be transferred into a Teflon®-sealed, screw-cap bottle and stored with minimal

Table 4-2 Solvents recommended for the preparation of headspace standards*

Solvent	Boiling point (°C)	Density, 20 °C/4 °C	Miscibility with water
Benzyl alcohol	205	1.045–1.046	40 g/L (19 °C)
Benzoic acid ethyl ester	214	1.046–1.047	0.5 g/L (20 °C)
Benzoic acid benzyl ester	324	1.117–1.119	
N,N-Dimethylacetamide	165	0.940–0.942	∞
N,N-Dimethylformamide	155	0.948–0.949	∞
Methyl cellosolve	124	0.964–0.965	∞
Ethyl cellosolve	135	0.929–0.930	∞
Propylene glycol carbonate	241–242	1.204–1.205	214 g/L (20 °C)
Glycerol	290	1.259–1.262	∞
1,3-Dimethyl-2-imidazolidone*	104 (15 torr)	1.055–1.057	∞

*Reference 25.
Source: Reference 24 except as indicated.

headspace in a freezer, protected from light. Unused standards may be stored in a freezer in sealed vials for several months, but if the vial is opened to take out an aliquot, the remaining solution should be used within a few days after opening. Aliquots are to be taken with help of a microsyringe.

If lower concentrations are required for calibration, an appropriate volume of the original stock solution should be transferred into another ground-glass, stoppered volumetric flask, which is then filled up to volume with the solvent. It is not necessary to use the same solvent in the secondary standard and in the first standard solution. For example, the best solvent for dissolving the component(s) of the standard might not be ideal from a chromatographic point of view because, say, of the presence of impurities. In such a case, an aliquot of the original stock solution can be further diluted by a more suitable solvent, reducing the possibility of interference by the impurities (which are diluted also).

Example 4.1 describes the preparation of such diluted stock solution to be used for the determination of volatile aromatic hydrocarbons in the atmosphere of an underground garage (see Example 5.10).

Example 4.1

Into a 10 mL volumetric flask, the following volumes of the aromatics are added:

Stock solution I

benzene	10 μL (0.879 mg/mL)
toluene	20 μL (1.734 mg/mL)
ethylbenzene	10 μL (0.867 mg/mL)
m-xylene	10 μL (0.864 mg/mL)
p-xylene	10 μL (0.861 mg/mL)
o-xylene	10 μL (0.880 mg/mL)

Then the flask is filled up to volume with propylene glycol carbonate (PGC); the concentration of this stock solution I is given in parentheses in the preceding list.*

From stock solution I a 100-μL aliquot is taken out with a microsyringe and added into another 10 mL volumetric flask, where it is filled up to volume with PGC. The final concentration of this stock solution (II) is:

Stock solution II	
benzene	8.79 μg/mL
toluene	17.34 μg/mL
ethylbenzene	8.67 μg/mL
m-xylene	8.64 μg/mL
p-xylene	8.61 μg/mL
o-xylene	8.80 μg/mL

In the case of volatile analytes and their solutions, we prefer volumetric additions rather than weighing. Volumetric additions are faster and less prone to sample losses by vaporization. Syringes are available in a wide volume range, and — if operated at room temperature and normal pressure — their accuracy and precision is very good (<1%). If needed, the added amount can be calculated considering the density of the substance, and it can also be controlled by weighing the closed and crimped vial.

As mentioned, vials containing the standard solutions can be stored in a freezer. When removed from the freezer, each vial should be warmed to 20 °C before an aliquot is taken out. This is because the volumes of the sampling devices (pipettes and syringes) are calibrated at 20 °C; therefore, we do not recommend that (empty) headspace vials and pipettes be chilled prior to being used to transfer samples from the original container as suggested by Penton [26].

We once observed a typical case of inaccuracies induced by temperature changes when a cold vial with a standard solution was taken from a refrigerator and immediately used to spike a series of headspace vials, by pipetting aliquots. There was a significant drift in the chromatographic results because between the spiking of the first vial with the still cold solution and the spiking of the last one, the standard had warmed to room temperature, thus changing steadily the actual amount of the added aliquots.

One important question associated with the preparation of such standard solutions and their use in quantitative analysis is the influence of the matrix of the sample. This is not restricted to standards: the matrix also influences the sample-to-sample reproducibility of quantitative analysis. Therefore, we shall discuss it in detail.

---

* The densities of the standard compounds used here are (in g/mL, 20 °C/4 °C): benzene = 0.879, toluene = 0.867, ethylbenzene = 0.867, m-xylene = 0.864, p-xylene = 0.861, o-xylene = 0.880.

4.4.2 Preparation of a Standard Solution from a Gaseous Compound

The preparation of a standard solution from a liquid or solid analyte can be carried out in the above-described way, by accurately measuring the weight or volume of the substance. In the case of gaseous substances, the procedure has to be properly adapted to the different characteristics of the material.

The cylinder containing the pure gaseous substance (e.g., ethylene oxide: Figure 4-7) is fitted with a syringe adapter, the needle of a gas-tight syringe is inserted into the adapter's septum, and the cylinder valve is opened. First the syringe barrel must be purged with the gas to replace the air, and for this step the plunger is removed. After a purge of a few seconds, the plunger is reinserted and the valve closed (Figure 4-7A). The measured gas volume is then transferred into a headspace vial closed with a septum and an aluminum crimp cap, which has been filled nearly completely with a suitable solvent, except a small remaining gas bubble. The actual amount of the solvent present can be determined by weighing. The gas volume in the syringe should be transferred into the vial by piercing the septum in such a way that the tip of the syringe needle ends in the gas bubble and does not touch the solvent (Figure 4-7B) because otherwise the good solubility of the gas may cause the solvent to be sucked back rapidly into the syringe barrel.

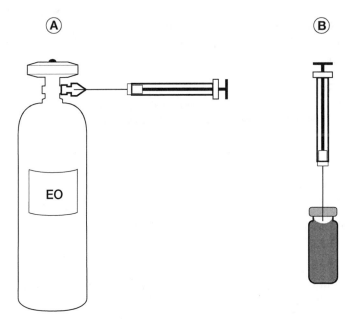

Figure 4-7. Preparation of a standard ethylene oxide solution. (A) Filling the gas-tight syringe with the gas from the cylinder. (B) Transferring the gas into a sealed headspace vial filled with water.

Since the measured gas volume in the syringe should be at atmospheric pressure, a gas syringe having the needle open to atmosphere can be used here to ensure that the amount is determined from the mole volume of the gaseous compound.

The amount of the transferred gas can be calculated with help of the general gas laws in two ways, considering the g-mole volume (V_{mole}) of the gas under standard conditions [22.414 L at 0°C and 101.08 kPa (760 torr)], as well as the atmospheric pressure (p_a) and ambient temperature (T_a, deg K). In the first case, the volume of the gas under given conditions is calculated directly from the fundamental relationship

$$p \cdot V = n \cdot R \cdot T \tag{4.5}$$

For one mole, $V = V_{mole}$ and $n = 1$; thus for atmospheric pressure and temperature

$$V_{mole} = \frac{R \cdot T_a}{p_a} \tag{4.6}$$

The value of gas constant R depends on the units used for pressure and the units in which V_{mole} should be expressed. If the volume is given in liters, the corresponding values are:

$$R = 0.08204 \, \text{L} \cdot \text{atm/deg} = 62.3 \, \text{L} \cdot \text{torr/deg} = 8.3127 \, \text{L} \cdot \text{kPa/deg}$$

From the mole volume V_{mole} and the actual volume V_i of the gas and its molecular mass, the transferred amount can be calculated. Examples 4.2 illustrate this calculation when preparing a standard solution of ethylene oxide in water.

Example 4.2′

First 5.0 mL of pure ethylene oxide (EO) gas was transferred from a gas cylinder with a gas-tight syringe into a closed small headspace vial (Figure 4-7) that already contained 8.75 g of water. Atmospheric conditions were as follows: $p_a = 100.8$ kPa; $T_a = 21 \, °C = 294.16$ K. The molecular weight of EO is 44.05. Its mole volume under the given conditions is:

$$V_{mole} = \frac{8.3127 \times 294.16}{100.81} = 24.256 \, \text{L}$$

The amount corresponding to 5.0 mL gas is:

$$\frac{5.0 \times 44.05}{24.256} = 9.080 \, \text{mg}$$

Thus, the EO concentration of the standard solution is 9.08 mg/8.75 g or 1.038 mg/mL (assuming that 1 mL = 1 g).

The second way of calculation first utilizes the general pressure–volume relationship:

$$p_1 \cdot V_1 = p_2 \cdot V_2 \tag{4.7}$$

In this way, the mole volume at p_a can be calculated, then corrected to the actual temperature T_a:

$$V_{pa,Ta} = \frac{p_o \cdot V_o}{p_a} \cdot \frac{T_a}{273.16} \qquad (4.8)$$

Example 4.2″

Using the preceding values, $V_o = 22.414$ L, $p_o = 101.08$ kPa, $p_a = 100.814$ kPa, $T_a = 294.16$ K:

$$V_{pa,Ta} = \frac{101.080 \times 22.414}{100.814} \cdot \frac{294.16}{273.16} = 24.201$$

Thus the amount corresponding to 5.0 mL gas is:

$$\frac{5.0 \times 44.05}{24.201} = 9.10 \text{ mg}$$

4.5 Influence of the Matrix

In this book we frequently refer to the *matrix*. This expression is defined by the dictionaries as the "enveloping element" within which something is enclosed or embedded. In headspace analysis we use the term to express the bulk of the sample that contains the volatile compounds to be measured. Usually the matrix is not a pure compound, but a complex mixture of compounds, some of which may be nonvolatile; but some may be volatile, although less than the analyte of interest. For example, in the determination of the blood alcohol level, the blood itself is the matrix. The interaction of the matrix components with the analyte influences its solubility (partition coefficient). This is what we call the *matrix effect*. In earlier discussions (e.g., Section 2.3.4), examples were given for this (e.g., the concentration of the analyte in the headspace of a pure aqueous solution will be quite different after equilibrium and in an aqueous solution with the same original analyte concentration, but also containing dissolved salts). Similarly, if the matrix is a mixture of two (or more) solvents, the distribution of the analyte between the two phases will depend on the quantitative composition of the matrix.

The matrix effect represents an important consideration in sample-to-sample reproducibility and particularly in the preparation of standard mixtures used for calibration or as external standards: it is obvious that (except in MHE measurements) these must have a matrix reproducing the actual sample matrix as closely as possible.

4.5.1 Clean Matrix Is Available

When a clean matrix is available, there is no problem in preparing a standard. This is the case when, for example, HS-GC is used to measure a contaminant that is not present in the usual samples. The determination of volatile aromatic hydrocarbons in used engine oil or traces of tetrachloroethylene in olive oil are typical examples and will be discussed in Chapter 5 (Figures 5.5 and 5.6).

Sometimes a "pure" matrix (i.e., one containing zero amount of the analyte) can be prepared from the natural sample by stripping off the volatile compounds using the MHE procedure. In Chapter 5 we will demonstrate this possibility (see Example 5.2). However, it must not be overlooked that some major volatile sample constituents — particularly water — will also be stripped off and thus, without compensating for these, the "pure" matrix prepared in this way will have a matrix effect different from that of the the actual sample.

4.5.2 Matrix Effect Can Be Eliminated

Sometimes the composition of the matrix is not known exactly, but it can be modified so that variation in the sample matrix will have no further influence on analyte distribution. Typical examples are the dilution of a liquid sample with a solvent and, in the case of the solution approach (Section 4.2), preparation of a diluted solution. Practical experience has shown that in general, the influence of matrix components present in concentrations less than about 1% can be ignored. Thus, by the proper dilution of the sample, we can sufficiently lower the concentration of the matrix components. The modified method for blood alcohol analysis discussed later (Section 5.2) is a typical example for this: by diluting the blood sample 1:10 with an aqueous solution of the internal standard, the effect of small variations in the matrix composition is eliminated.

Elimination of the matrix effect by sufficient dilution is possible in the case of dissolved inorganic salts: it has been demonstrated that salt concentrations up to about 5% can be ignored. This is shown in Table 4-3, reporting on the determination of several halogenated hydrocarbons in water, with different salt (Na_2SO_4) concentrations, starting with pure water. In the case of significant matrix influence, differing peak area values would be expected. However, the peak areas listed in Table 4-3 show only random variations within the usual headspace precision. These results have important implications in the determination of volatile organic compounds (VOCs) in water by HS-GC, which is an important application of the technique.

When considering the elimination of the matrix effect by dilution, one should never forget that dilution automatically reduces the headspace sensitivity.

The matrix effect can also be eliminated by completely evaporating the analyte in the vial. In this case, there is no longer any partitioning between two phases. These techniques are discussed later, in Section 4.6.

Although inorganic salt concentrations up to about 5% may be ignored, any salt content above this range will have significant influence on the results. To demonstrate this, we prepared an aqueous solution of trichloroethylene (TCE) with a concentration of 0.26 μg/mL (260 ppb) and added equal volumes of this

Table 4-3   Peak area values obtained in the headspace analysis of aqueous solutions of halogenated hydrocarbons,* containing various concentrations of Na_2SO_4

Analyte	Concentration ($\mu g/L$; ppb)	Peak area (counts) at four Na_2SO_4 concentrations				Mean peak area (counts)	RSD ($\pm$%)
		0.0%	0.02%	1.0%	5.0%		
Dichloromethane	3750	279	278	278	278	278	0.18
Chloroform	295	808	808	787	779	796	1.86
Trichloroethylene	260	1574	1569	1506	1507	1539	2.44
Carbon tetrachloride	65	1647	1621	1540	1547	1600	3.32
Dichlorobromomethane	80	1119	1104	1061	1043	1082	3.30

*Sample: 5 mL, equilibrated at 80 °C for 60 minutes; electron capture detector.

sample (5 mL) into two headspace vials. To one of the solutions we also added Na_2SO_4, at a concentration of 20%. After equilibration of the two samples at 80 °C for 30 minutes, using a shaker, both samples were analyzed under identical conditions, and the following peak area values were obtained:

TCE solution, no salt	1738 counts
TCE solution, 20% Na_2SO_4	2073 counts

In other words, the peak area obtained from the salt solution was 19.3% higher than obtained for the pure aqueous solution. The reason for this can be explained by the general theory of headspace analysis: the addition of salt reduces the solubility, hence the partition coefficient, of TCE in water, with the concomitant increase of its concentration in the headspace of the vial. However, an additional volume effect due to the added amount (volume) of the salt and its influence on the phase ratio β and thus on the sensitivity was already discussed in Section 2.3.4.

It should be mentioned that if we were to carry out exhaustive gas extraction of these two samples, using the multiple headspace extraction process (see Section 5.5), the total peak area (which is now proportional to the total amount of analyte present, without any matrix influence) would be practically the same for both of them. The results of using a five-point MHE measurement were as follows:

TCE solution, no salt	3546 counts
TCE solution, 20% Na_2SO_4	3647 counts

The difference between the pure aqueous and the salt solution is now only 2.8%.

4.5.3 Artificial Matrix Can Be Prepared

When the concentration of the main components of the matrix is known, one can prepare an artificial matrix, ignoring the minor components. A typical example is the investigation of certain components in alcoholic beverages. Since the concentrations of ethanol and water in the beverage are known, one can prepare an aqueous ethanol solution with the same ethanol concentration and use it as a *simulated matrix*. An example will be shown in Figure 5-8 illustrating the determination of methanol in adulterated wine.

4.6 Methods Aiming at the Complete Evaporation of the Analyte

One way to eliminate the matrix effect is to completely evaporate the analyte in the sample vial. This can be achieved by sufficiently reducing the sample volume and using an appropriate thermostatting temperature. This can be explained by reconsidering eq. 2.12:

$$K = \frac{W_S}{W_G} \cdot \beta \tag{2.12}$$

where W_S and W_G are the respective amounts of the analyte in the sample and gas phases upon equilibrium, and β is the phase ratio of the vial, the ratio of the volumes of the headspace (V_G) and of the sample (V_S):

$$\beta = \frac{V_G}{V_S} = \frac{V_V - V_S}{V_S} \tag{2.3}$$

where V_V is the volume of the vial. A significant reduction in the sample volume (down to a few microliters) will result in a corresponding increase* of the phase ratio. The partition coefficient is constant at a given temperature; therefore, if the phase ratio is increased, the value of W_S/W_G will be simultaneously decreased, by the same factor. Since ($W_S + W_G$) is constant, this can be achieved only by an increase in W_G and a concomitant decrease in W_S, which will approach zero. An increase in the temperature reducing the partition coefficient could further accelerate this process.

We can divide the methods aiming at the full evaporation of the analyte and/or the whole sample into two groups.

4.6.1 The Total Vaporization Technique (TVT)

When the whole sample, including the components of the matrix, will evaporate, we no longer have a headspace and a condensed phase: only a single phase is present in the vial, a gas (vapor), and the vial only serves as a container for

---

* Using a vial with a volume of 22.3 mL, if the sample volume is 2 mL, $\beta = 10.15$. On the other hand, with a sample volume of 10 μL, $\beta = 2229$, representing a 220-fold increase.

it. Thus, the concentration of the analyte in the vial is:

$$C_G = W_o/V_V \qquad (4.9)$$

where V_V is the vial's volume and W_o is the amount of analyte present in the original sample. We call this variant the *total vaporization technique* (*TVT*).

The TVT is particularly suited for the preparation of vapor standards to be used for calibration. This in situ preparation of the gaseous standard allows analysts to avoid the problems well known with low-concentration standard gas mixtures stored in glass or metal containers. In addition, since the vial can be heated to higher temperatures, it is also feasible to prepare vapor standards from substances that are liquids at room temperature by vaporizing a few microliters of the compound or its diluted solution in the vial. By the proper selection of the analyte concentration in the solution injected into the vial, gas concentrations in the ppb range can also be obtained.

It is interesting to note that although one would expect the total vaporization to occur instantaneously at the proper elevated temperatures (comparable to instantaneous vaporization in the GC injector upon injecting a liquid sample with a syringe), this is not always so. It has been experienced that evaporation under static conditions in the vial takes some time, even if the vial temperature is above the boiling point of the sample constituents. For this reason, a certain "equilibration time" is also needed when the total vaporization technique is utilized.

Naturally, the amount of the liquid sample that can be evaporated is limited, since the concentration in the gas must not exceed that of a saturated vapor: otherwise a condensed phase will also be formed.

With regard to the admissible sample volume for the TVT, a rough estimate can be made in the following way. The mole volume of a gas [at 0 °C and 101.3 kPa (1 atm) pressure] is 22.4 L, while the volume of the standard headspace vial used by us is 22.3 mL. This means that a vial accepts approximately 1 millimole of a sample. In the case of an aqueous solution, this corresponds to 18 mg (18 µL). Because of expansion of vapor at elevated temperatures, however, and because the air in the vial already contains some water due to its humidity,* one should reduce this volume by about 30%. Thus, the volume of an aqueous solution should not exceed about 13–15 µL if the TVT is to be utilized. This rough estimate — 70% of the volume corresponding to one millimole of the solvent — is recommended in general.

A more accurate estimation of the sample volume used in TVT could be made by using the basic gas laws, but even that will not be precise, since the sample is mostly a mixture and not a pure compound.

In the discussion of the various quantitative methods (see Chapter 5) the TVT will be used in a number of examples for the preparation of the external (vapor) standard.

---

* The density of saturated water at 20 °C is 0.0173 mg/mL (cf. Table 2-1). Thus an empty vial of 22.3 mL volume will contain at 60% humidity $(22.3 \times 0.0173) \times 0.6 = 0.231$ mg of water.

4.6.2 The Full Evaporation Technique (FET)

Apparently the matrix no longer influences the phase distribution if the analyte is completely transferred from the sample into the gas phase. In practice this is realized by adjusting the phase ratio β as high as possible. Since the volume of the vial is limited, the sample volume must be decreased: thus a small sample is extracted by a relatively large gas volume, and under favourable conditions an exhaustive extraction can be achieved. Markelov and Guzowski [27] have called this variant the *full evaporation technique (FET)*. Contrary to the foregoing TVT, some components of the matrix still remain in condensed state. However, this residue in the vial has (or should have) no further influence on the analyte's distribution in the vial. In other words, we may still speak about "headspace analysis," since we have two phases in the vial, although now all the analyte molecules should be in the headspace, with the analyte concentration in the remaining sample approaching zero.

Since we no longer have a distribution of the analyte molecules, $K \to 0$, and the concentration of the analyte in the vial's headspace will be:

$$C_G = \frac{W_o}{V_V - V_S} \tag{4.10}$$

where W_o is the total amount of the analyte present, V_V is the vial's volume, and V_S is the volume of the remaining sample phase in the vial. It is the intended purpose of the FET to have the volume of the remaining nonvolatiles become negligibly small and eq. 4.10 become identical to eq. 4.9, thus approaching the ideal relationship of the TVT.

As mentioned, the FET assumes full evaporation of the analyte. Whether this is really achieved depends on both the phase ratio and the partition coefficient, a relationship that can be expressed by the *extraction yield*, which should always be better than 90% (preferably >95%). Next we describe the calculation of the extraction yield, and Table 4-4 gives data for various sample volumes (phase ratio values) and partition coefficients. As seen for high values of K, only very small sample volumes can be used: for example, at $K = 500$ (which is not an unusual value), even a volume as small as 20 µL would only result in an extraction yield of 69%. At low values of K (i.e., in the case of analytes with low solubility in the matrix), one may use larger sample volumes; however, it is important that the analyte concentration in the headspace does not exceed that of a saturated vapor: otherwise part of it will remain dissolved in the matrix.

The FET therefore is preferably applied when the sample matrix will also evaporate (e.g., for aqueous samples), thus avoiding the formation of any condensed phase, while the nonvolatile material remains as a dry, inert solid residue with no adsorptive properties (cf. Section 6, Figure 6-6).

Another limitation of the FET is headspace sensitivity. As discussed shortly (Section 4.6.4), regular headspace analysis generally has a higher sensitivity than FET or TVT, except when partition coefficient values are very high.

A practical example of the FET was given by Kolb [28,29], who described a procedure for the quantitative analysis of propellant gases and volatile solvents present in aerosol packings. An aliquot of the liquefied homogeneous sample was transferred into a headspace vial: the volatile compounds could fully evaporate there, while the nonvolatile sample constituents such as surfactants, resins, and other ingredients remained as a residue in the vial. Another example for use of the FET is the determination of monomers and residual solvents in a polymer dispersion. The problem of using standard HS-GC is that such a dispersion represents a three-phase system (polymer droplets + water phase + gas phase) in the vial, and it is difficult to prepare an external standard for calibration consisting of an aqueous dispersion without any monomer in it. The classical technique, therefore, would be the solution approach: for example, the dispersion may be dissolved in dimethylformamide and the homogeneous solution treated just as a liquid sample [7]. This procedure works well for concentrations above 100 ppm; however, with the more recent demand to determine lower concentrations, the dilution associated with the solution approach is no longer acceptable.

Two alternatives are possible: one of them is the FET technique, and the other one is to use the dispersion without any dissolution (see Section 4.1.3 and Figure 4-4) and to calibrate by the technique of standard addition, which, however, is discussed later (Section 5.4). With the FET approach, the three-phase system is destroyed by evaporation of the aqueous phase including all volatile analytes at 130 °C, while the polymer remains as a solid phase in the vial. A typical chromatogram for such an application is shown in Figure 4-8: here a polymer dispersion of only 20 mg was placed into the headspace vial and the volatile components plus water were evaporated during thermostatting at 130 °C.

A particular problem that must be considered with the FET is whether the remaining residues in the vial cause some adsorption effects. This can be tested by using multiple headspace extraction. If we prepare an external vapor standard by TVT and analyze it and the sample under identical conditions by MHE, parallel plots (see later: case E in Figure 5-15) indicate that they represent the same state. That is, the analytes in the sample are fully evaporated without any residual effects, just as in the case of the external vapor standard. If this is proven, the analysis of such systems can be carried out using external vapor standards as described in the appropriate place (Section 5.3). On the other hand, if the results of sample analysis show a nonlinear behavior (case F in Figure 5-15), a secondary effect of the suspended particles is indicated. An example of such adsorption effects by the residues in the vial is presented in Chapter 6 (cf. Figure 6-6).

4.6.3 Calculation of the Extraction Yield in FET

The extraction yield $Y\%$ of the full evaporation technique is the ratio of the amount of the analyte in the gas phase W_G to its total amount in the original

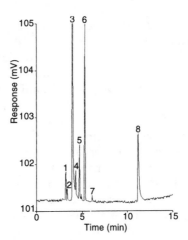

Figure 4-8. Determination of the monomers and residual solvents in an aqueous dispersion by FET. *HS conditions*: Sample: 20 mg, equilibration at 130 °C for 30 minutes. *GC conditions*: 50 m × 0.32 mm I.D. fused-silica, open-tubular column, coated with cyanopropyl (7%), phenyl (7%), and methyl (86%) silicone stationary phase; film thickness: 1 μm. Column temperature: isothermal at 70 °C for 8 minutes, then programmed at 10 °C/min to 90 °C. Split sampling. Flame-ionization detector. *Peaks* (concentration): *1* = acetaldehyde (16.2 μg/g), *2* = vinyl chloride (14.5 μg/g), *3* = ethanol (251 μg/g), *4* = isopropyl alcohol (12.7 μg/g), *5* = *tert*-butyl alcohol (44.2 μg/g), *6* = vinyl acetate (151 μg/g), *7* = ethyl acetate (6.8 μg/g), *8* = butyl acrylate (96 μg/g).

sample W_o, multiplied by 100:

$$Y\% = \frac{W_G}{W_o} \cdot 100 \tag{4.11}$$

Equation 2.12 expressed the partition coefficient as

$$K = \frac{W_S}{W_G} \cdot \beta \tag{2.12}$$

where W_S and W_G are the respective amounts of the analyte in the sample phase and headspace at equilibrium. Since $W_S + W_G = W_o$, we can write eq. 2.12 as:

$$K = \frac{W_o - W_G}{W_G} \cdot \beta \tag{4.12}$$

and

$$W_o = W_G \cdot \left(\frac{K}{\beta} + 1 \right) \tag{4.13}$$

Substituting to eq. 4.11 gives:

$$Y\% = \frac{100}{(K/\beta) + 1} \tag{4.14}$$

Table 4-4 lists the results of such calculations for partition coefficients ranging from 0.1 to 500 and sample volumes between 5 μL and 10 mL, considering a headspace vial having a volume of 22.3 mL. The phase ratio values were

Table 4-4 Extraction yield (%) of the FET for various values
of K and β ($V_{vial} = 22.3$ mL)

Sample volume:	10 mL	5 mL	2 mL	1 mL	100 μL	50 μL	20 μL	10 μL	5 μL
Phase ratio, β:	1.23	3.46	10.15	21.3	222	445	1114	4459	2229
K									
0.1	92.48	97.19	99.02	99.53	99.95	99.98	99.99	99.99	99.99
1.0	55.16	77.58	91.03	95.52	99.55	99.78	99.91	99.96	99.98
10	10.95	25.71	50.37	65.01	95.57	97.80	99.11	99.55	99.78
100	1.22	3.34	9.21	17.56	68.94	81.65	91.76	95.711	97.81
500	0.25	0.69	1.99	4.08	30.74	47.09	69.02	81.68	89.92

calculated according to the usual relationship, given earlier in eqs. 2.2 and 2.3.

$$(2.3)\quad \beta = \frac{V_G}{V_S} = \frac{V_V - V_S}{V_S}$$

where V_V, V_G, and V_S are the respective volumes of the vial and of the gas and sample phases in it.

4.6.4 Comparison of Headspace Sensitivities

It is intuitive to compare the sensitivity of determinations using the TVT or FET to results obtained by means of standard headspace techniques. For this comparison let us consider an aqueous sample containing 1 ppm (1000 ng/mL) concentration of the analyte.

First, for TVT or FET, we added 15 μL of the aqueous solution into the headspace of the vial of 22.3 mL volume. The 15 μL liquid contains 15 ng of the analyte, and its concentration in the vial's gas volume will be $15/22.3 = 0.673$ ng/mL.

Now, for standard HS analysis, let us assume that we added 5 mL of the 1 ppm solution into a vial having a volume of 22.3 mL. The concentration of the analyte in the headspace C_G will depend on its original concentration in the sample c $C_o = 1000$ ng/mL), the phase ratio ($\beta = 3.46$), and the partition coefficient K, according to the basic relationship of headspace analysis:

$$C_G = \frac{C_o}{K + \beta} \qquad (2.19)$$

Let us consider five values for the partition coefficient; the calculated values of C_G will be as follows:

K	C_G (ng/mL)
1	224
10	74.3
100	9.7
500	2.0
1000	1.0

If we compare these results with the gas phase concentration $C_G = 0.673$, obtained by TVT or FET, we can conclude that regular HS-GC analysis could always result in a higher sensitivity than the methods using complete evaporation of the analyte (TVT or FET): the sensitivities of the two techniques approach each other only in the case of extremely high partition coefficient values. Thus, in general, the main reason for using the methods of complete evaporation is not to gain in sensitivity. Rather, one is interested either in eliminating the matrix effect (and thus the accompanying calibration problems for quantitative anaylsis) or having the advantage of the ease of preparing vapor standards with known (and very low) analyte concentration.

References

1. L. S. Ettre, C. Welter, and B. Kolb, *Chromatographia*, **35**, 73–84 (1993).
2. B. Kolb, C. Welter, and C. Bichler, *Chromatographia*, **34**, 235–240 (1992).
3. B. Kolb, P. Pospisil, and M. Auer, *Chromatographia*, **19**, 113–122 (1984).
4. *ASTM E 1142-93b*: Standard Terminology Relating to Thermophysical Properties.
5. M. Hoffmann, H. Kröner, and R. Kuhn, *Polymeranalytik*, *Vol. I*, Georg Thieme. Verlag, Stuttgart, 1977.
6. W. P. Brennan, *Thermal Analysis Study No. 7*, Perkin-Elmer Corporation, Norwalk, CT, 1973.
7. H. Hachenberg, and A. P. Schmidt, *Gas Chromatographic Headspace Analysis*, Heyden & Son, London, 1977; pp. 47–50.
8. S. J. Romano, *Anal. Chem.* **45**, 2327–2330 (1973).
9. A. R. Berens, L. B. Crider, C. J. Tomanek, and J. M. Whitney, *J. Appl. Polym. Sci.* **19**, 3169 (1975).
10. *ASTM D 4740-93*: Standard Test Method for Residual Vinyl Chloride Monomer in Poly(vinyl chloride) Resins by Gas Chromatographic Headspace Technique.
11. L. Rohrschneider, *Z. Anal. Chem.* **255**, 345–350 (1971).
12. *U.S. Pharmacopeia XXIII*. Organic Volatile Impurities (467), Method IV, 1995; pp. 1746–1747
13. R. J. Steichen, *Anal. Chem.* **48**, 1398–1402 (1976).
14. *ASTM D 4322–83(91)*: Standard Test Method for Residual Acrylonitrile Monomer in Styrene–Acrylonitrile Copolymers, and Nitrile Rubber by Headspace–Gas Chromatography.
15. *ASTM D 4443–84(89)*: Standard Test Method of Analysis for Determining the Residual Vinyl Chloride Monomer Content in ppb Range in Vinyl Chloride Homo, and Copolymers by Headspace–Gas Chromatography.
16. B. Kolb, *LC/GC Int.* **8**, 512–524 (1995).
17. J. P. Hsu, G. Miller, and V. Moran III, *J. Chromatogr. Sci.* **29**, 83–88 (1991).
18. G. Machata, *Arbeitsmed. Sozialmed. Präventivmed.* **21**, 5–7 (1986).
19. B. L. Karger, L. R. Snyder, and C. Horváth, *An Introduction to Separation Science*, Wiley, New York, 1973; pp. 67–68.
20. M. Dong, A. H. DiEdwardo, and F. Zitomer, *J. Chromatogr. Sci.* **18**, 242–246 (1980).
21. Courtesy of F. Kurt Retsch GmbH & Co., D-42759 Haan, Germany.
22. *ASTM F151–86(91)*: Standard Test Method for Residual Solvents in Flexible Barrier Material.
23. Methode zur Bestimmung von Restlösemitteln in lackierten Aluminiumfolien. Merkblatt 57, *Verpack. Rundsch.* **7**, 56–57 (1989).
24. Merck-Schuchard, *Chemikalien zur Synthese, Manual 94/96*, Dr. Theodor Schuchard & Co., Hohenbrunn, Germany, 1994.
25. M. DeSmet, K. Roets, L. Vanhoof, and W. Launvers, *Pharmacopeial Forum*, **21**, 501–514 (1995).
26. Z. Penton, *HRC*, **15**, 834–836 (1992).
27. M. Markelov, and J. P. Guzowski, *Anal. Chim. Acta*, **276**, 235–245 (1993).
28. B. Kolb, *Aerosol Report*, **24**, 619–632 (1985).
29. B. Kolb, *Aerosol Age*, pp. 42–62, April 1986.

Headspace Methods
for Quantitative Analysis

Although headspace–gas chromatography may also be used in qualitative analysis (see Chapter 8), its main application is obviously for the quantitative determination of volatile compounds present in liquid, solid, or gaseous samples, and this is carried out by analyzing an aliquot of the headspace of the sample vial. Since the actual determination is carried out by gas chromatography, the general techniques used in GC for quantitative analysis also apply to HS-GC. Thus, we can evaluate the chromatogram by *internal normalization* or by using *internal* or *external standards*. In addition, there are two further methods specially used in HS-GC. In the first, referred to as *standard addition*, known amounts of the analyte are added to the actual sample, and the original analyte concentration (amount) is established from the increase in the peak area upon addition of a known amount. The second special method is the so-called *multiple headspace extraction* (MHE). This resembles total gas extraction; however, it is carried out stepwise, and only in a few steps: the total amount of the analyte present is then calculated by utilizing the well-established relationships of a geometric progression.

As we learned in Chapter 4, the total amount of a highly volatile analyte present in the sample may be evaporated. In this case, the headspace contains all the analyte. This is the *total vaporization technique* (TVT) or the *full evaporation technique* (FET). Finally, the headspace vial may also act as a sampler for gas samples. In these three cases, all the analyte molecules are present in the gas phase of the vial and no partitioning is involved in the determination. Therefore, even the technique of normalization can be applied, which otherwise does not make much sense in a real headspace analysis.

Generally, we assume that an *equilibrium* is reached in the sample vial prior to the actual transfer of an aliquot of the headspace into the column. In some cases, however, analysis is carried out before equilibrium is reached; *nonequilibrium analysis* applications are discussed in Chapter 7.

Quantitative analysis in GC always depends on *calibration*, and when an external standard or the method of standard addition is used, it is based on the comparison of two independently obtained chromatograms. In such cases, assurance of sampling reproducibility is always an important question. Since the widely used, manually operated microsyringes had relatively poor reproducibility, the internal standard technique was the preferred method in gas chromatography, and this philosophy was inherited by headspace GC. With the presently available automated sampling systems, including automated headspace samplers with their high precision, the other calibration techniques should be applied more frequently. However, for HS-GC, the term "sampling precision" needs to be clarified.

In HS-GC, the transfer of an aliquot of the headspace gas from the vial into the column is performed with high precision; therefore, the main questions are how reproducibly the sample itself must be placed into the headspace vial and how any variation of the sample volume (phase ratio β) will influence the overall precision. We have already discussed this complex relationship in Section 2.3: as explained there, small differences in the sample volume may be neglected in the case of analytes having high partition coefficients (i.e., a high solubility in the sample matrix). On the other hand, compounds with small partition coefficients (high volatility and relatively poor solubility) require exact reproducibility of the sample volume for calibration and analysis. Since in general, the partition coefficient of the analyte is not known, it is always advisable to use the same sample volume in quantitative analysis.

The subsequent chapters outline the individual techniques used for quantitative analysis; in this discussion, we shall generally speak about the "amount" of the analyte or of the standard. Depending on the method of calibration, however, "amounts" may also mean concentration; and the latter may be expressed on a weight/weight or a weight/volume basis. Similarly, we shall generally speak about the peak area A; however, in certain circumstances, one may also use peak heights. The term "sample" generally refers to the original sample and subscript o indicates data related to this. Data related to the standard are usually indicated by the subscript st, or in the case of an external standard, ex.

5.1 Internal Normalization

The general purpose of headspace analysis is the determination of the amount (concentration) of some of the volatile components of a sample, which also contains lowvolatile or nonvolatile components representing the matrix. Thus, internal normalization—where the total composition of the sample is established—makes hardly any sense, except in two cases: when only the gas phase composition is of interest, and when all the volatiles are completely vaporized into the gas phase of the vial.

In internal normalization, the sum of the peak area is established and then the relative amount of each compound is calculated:

$$A_1 + A_2 + \cdots + A_i + \cdots + A_n = \sum_{i=1}^{n} A_i \tag{5.1}$$

$$A_i\% = \left(A_i \middle/ \sum_{i=1}^{n} A_i \right) \cdot 100 \tag{5.2}$$

Assuming that equal amounts (concentrations) of each analyte present would give the same peak area, one may take $A_i\%$ as concentration.

The following example (Figure 5-1 and Table 5-1) demonstrates the determination of light hydrocarbons present in an oil-well cutting from 4000 m depth. After the elution of the lighter hydrocarbons the column, was backflushed, obtaining the heavier hydrocarbons as a composite peak.

Table 5-1 Determination of light hydrocarbons in oil-well bore cuttings*

Peak no.	Retention time (min)	Compound	Peak area (counts)	Area (%)
1	1.51	Methane	21,640	0.217
2	1.90	Ethane	33,656	0.337
3	2.50	Propane	262,156	2.625
4	3.20	Isobutane	2,017,484	20.204
5	3.72	n-Butane	342,784	3.433
6	5.28	Isopentane	913,254	9.146
7	5.92	n-Pentane	364,518	3.651
8	8.50		638,412	6.393
9	9.25		218,969	2.193
10	9.93		327,769	3.183
11	11.81		75,091	0.752
12	12.16		135,692	1.359
13	14.38	Benzene	215,692	2.160
14	15.12		586,624	5.875
15	17.17		330,880	3.314
16	21.28			
17	22.82			
18	24.00	Backflushed	4,276,069	34.811
19	25.32			
			10,760,690	99.753

* For conditions, see the caption of Figure 5-1.

Source: Burton S. Todd, Perkin-Elmer Corporation, Norwalk, CT.

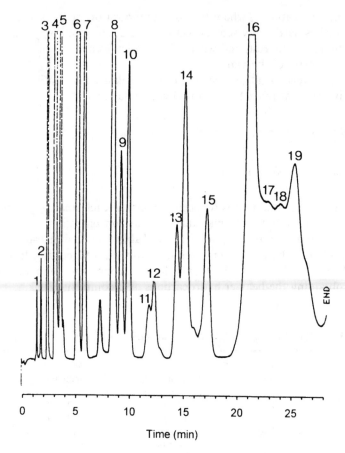

Figure 5-1. Analysis of an oil-well bore cutting. *HS conditions*: Sample: 2.0 g of oil-well cutting from 4000 m depth, equilibrated at 50 °C. *GC conditions*: Two 6 ft × 1/8 in. O.D. packed columns containing 30% DC-200/500 methylsilicone stationary phase on Chromosorb P 60/80 mesh; backflush mode. Column temperature: 120 °C. Flame-ionization detector. For peak identification, see Table 5-1.

Source: Burton S. Todd, Perkin-Elmer Corporation, Norwalk, CT.

In the preceding example, we assumed that the peak area percentages are equal to concentration. In exploratory work, or if the analytes are similar in chemical nature or have a higher molecular weight (chain length > about C_6), this assumption may be accepted. However, this is not always true. In many cases, the same amounts (concentrations) of different analytes give different peak areas. In other words, concentration is *proportional* to the peak area

$$C_i = f_i \cdot A_i \tag{5.3}$$

but the factor f_i is different for each compound present: it is a substance-specific calibration factor. We may call $f_i \cdot A_i$ the *reduced peak area*. Similarly to eq. 5.1

we can write

$$A_1 f_1 + A_2 f_2 + \cdots + A_i f_i + \cdots + A_n f_n = \sum_{i=1}^{n} A_i f_i \tag{5.4}$$

and

$$C_i\% = (A_i f_i)\% = \frac{A_i f_i}{\sum\limits_{i=1}^{n} A_i f_i} \cdot 100 \tag{5.5}$$

Only if the volatile analytes are completely vaporized (e.g., by TVT or FET) are these calibration factors f_i identical with the *response factors* (RF) which reflect the differences in the detector's response to various compounds. However, in normal HS-GC they also include differences in the partition coefficients of the volatile compounds present in a liquid or solid sample. Therefore, it is more appropriate here to call the factor f_i a *calibration factor*, to avoid any confusion with the RF values used in general GC.

The calibration factors can be determined by analyzing the mixture of two (or more) compounds of known concentrations (amounts). Assuming proportionality between concentration and peak area (eq. 5.3), we can write

$$C_1 = f_1 A_1$$
$$C_2 = f_2 A_2$$
$$\frac{C_1}{C_2} = \frac{f_1}{f_2} \cdot \frac{A_1}{A_2} \tag{5.6}$$

Since the calibration factors are relative values, one can assign the value of unity to one of them and express the others relative to it. We call this compound the *standard* and indicate it with the subscript *st*, while the other compound is indicated with subscript i. Thus, if $f_1 = f_{st} = 1.00$ and $f_2 = f_i$, then:

$$\frac{C_{st}}{C_i} = \frac{1.00}{f_i} \cdot \frac{A_{st}}{A_i}$$

and

$$f_i = \frac{C_i}{C_{st}} \cdot \frac{A_{st}}{A_i} \tag{5.7}$$

Example 5.1 illustrates this calculation, including the determination of the calibration factors. The sample was a polymer foam, and the task was to establish the concentration of the propellant gases. The full evaporation technique was used. The gas mixture contained three main components: isopentane, *n*-pentane, and 2-methylpentane, representing 99.2%; the other compounds were present in small concentrations. For this reason, calibration factors were determined only for the three main compounds: for the others it

Table 5-2 Determination of the calibration factors for three hydrocarbons
analyzed in Example 5.1

Compound	Volume (μL)	Density (g/mL)	Weight present (mg)	Peak area* (counts)	Calibration factor
n-Pentane	1.333	0.625	0.833	437,270	1.000
Isopentane	1.333	0.621	0.828	431,306	1.008
2-Methylpentane	1.333	0.672	0.896	454,453	1.035

* Mean of three parallel determinations.

was considered to be $f = 1.000$.

Example 5.1

Determination of the calibration factors. Equal volumes of three hydrocarbons were
mixed and 4.0 μL of this mixture was vaporized (TVT) at 80 °C in an empty vial,
and analyzed under the conditions that follow. The calibration factors were calculated using eq. 5.7 with n-pentane as the standard compound. Table 5-2 gives the
pertinent data.

Analysis of the sample. A 4.3 mg sample of the polymer was placed in the headspace
vial and thermostatted at 120 °C for 45 minutes, using FET. For the GC conditions,
see the caption of Figure 5-2.
   Table 5-3 gives the analytical results and the results of the calculation of the
composition, using eq. 5.5.

Table 5-3 Determination of the composition of a propellant gas mixture
(Example 5.1)

Compound	Peak area A_i (counts)	f_i	$f_i \cdot A_i$	Wt%
Isobutane	1,614	1.000*	1,614	0.05
n-Butane	1,809	1.000*	1,809	0.06
Isopentane	777,993	1.008	784,217	24.98
n-Pentane	2,276,277	1.000†	2,276,277	72.50
2,2-Dimethylbutane	4,912	1.000*	4,912	0.15
2,3-Dimethylbutane	14,684	1.000*	14,684	0.47
2-Methylpentane	51,640	1.035	53,447	1.70
3-Methylpentane	2,708	1.000*	2,708	0.09
			3,139,668	100.000

* Because of the very low concentration of this analyte, a response factor of $f = 1.000$ was assumed.
† Standard.

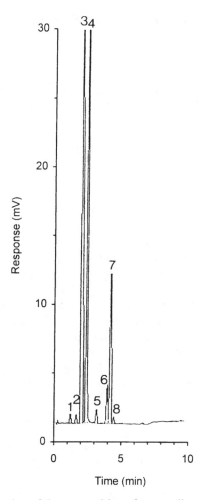

Figure 5-2. Determination of the composition of a propellant gas mixture. For details see Example 5.1. *HS conditions*: Sample: 4.3 mg polymer. Equilibration: 45 minutes at 120 °C. *GC conditions*: 50 m × 0.32 mm I.D. fused-silica, open-tubular column coated with bonded methyl silicone stationary phase; film thickness: 5 μm. Column temperature: 80 °C. Splitless sampling. Flame-ionization detector. *Peaks*: *1* = isobutane, *2* = *n*-butane, *3* = isopentane, *4* = *n*-pentane, *5* = 2,2-dimethylbutane, *6* = 2,3-dimethyl-butane, *7* = 2-methylpentane, *8* = 3-methylpentane. For quantitative data, see Table 5-3.

5.2 Internal Standard Method

In the internal standard method a standard compound of known amount (concentration) is added to the sample. As prerequisites, the standard must not be present in the sample, and there must be no compound present that has the same retention time in the chromatogram. From eq. 5.7 we can write

$$C_i = C_{st} \cdot f_i \cdot \frac{A_i}{A_{st}} \tag{5.8}$$

In serial determinations one can fix the concentration of the standard: thus, the value of $C_{st} \cdot f_i$ will be a constant. In this case, one may use this value as a combined *calibration factor*, f_c:

$$f_c = C_{st} \cdot f_i \tag{5.9}$$

and eq. 5.8 becomes:

$$C_i = f_c \cdot \frac{A}{A_{st}} \tag{5.10}$$

The advantages of the internal standard method are twofold compared to the other quantitative calculation methods. Only a single measurement is needed, as opposed to two or more; and, compared with the external standard method (see Section 5.3), the effect of small variations in the matrix can be eliminated by using a suitable compound for an internal standard. For example, in blood alcohol determination, the salt or lipid content of the blood may slightly vary: thus, if we use "standard blood samples" with added ethanol concentration as external standards, the composition of the actual blood sample (its salt or lipid content) may differ somewhat from that of the standard sample; therefore, the partition coefficient of ethanol may be slightly different in the two samples. On the other hand, if we apply the internal standard method, with another alcohol used as the internal standard, this alcohol suffers the same matrix effects as ethanol, due to their identical chemical polarity, and thus matrix differences are compensated. The situation is similar when otherwise similar samples would differ in their humidity: an internal standard compensates for such variations.

It is implied, then, that the compound used as the internal standard should have a polarity as similar to the analyte as possible. This is clearly different from the general use of internal standards in general GC, where any compound may be used as an internal standard, independent of its chemical properties.

In certain cases, particularly in official validation methods, one may rely on an external standard for quantitation, but still add an internal standard to the sample. In this case, any meaningful variation in the matrix would be observable in the relative size of the internal standard peak, and thus one can compensate for matrix differences if necessary.

While the advantage of the internal standard is that a single measurement only is required, let us not forget that the calibration factor still must be determined separately, representing additional measurements. Moreover, for the establishment of the calibration factor, one must reproduce the matrix, which may not be easy; one way to overcome this problem is to dilute the sample so much that the matrix effects are eliminated for all practical purposes.

A good example is the modified method for blood alcohol determination, discussed shortly.

A particular disadvantage of the internal standard method is that the addition of the standard to every sample is laborious and subject to the introduction of significant errors, particularly when the sample contains compounds of high volatility. Also in practical cases when the sample is filled into headspace vials in the field or at production facilities, often by unskilled workers, the vials should be closed as fast as possible and the addition of an internal standard may be unreliable. Therefore, the internal standard method is recommended only if other methods would create difficulties or would be too time-consuming.

Finally, one should remember that the addition of an internal standard assumes that in the chromatogram, there is sufficient space for the peak of that standard.

The example that follows for the internal standard method determines the residual amount of trichloroethylene in a gelatin capsule, using tetrachloroethylene as the internal standard. It also illustrates the establishment of the calibration factor. For this a capsule containing no trichloroethylene is needed to be able to reproduce the matrix. Since, however, such a "pure" capsule was not available, we prepared one by first putting a capsule through a multistep MHE procedure (see Section 5.5) until all the trichloroethylene content was purged away. Water is added to the capsule, which, at the thermostatting temperature (110 °C) will be dissolved in it: thus, we are analyzing a solution. There is, however, an additional reason for adding water. When trichloroethylene was stripped from the capsule by the MHE process, most of the water content was stripped, as well. Using this "dry" capsule would result in a matrix different from the actual capsule. Adding an excess of water to both calibration standard and sample eliminates this problem.

Example 5.2

Determination of the calibration factor. A pure gelatin capsule +1 mL water were added into the headspace vial and then 5 μL of a solution containing 0.5% each of trichloroethylene (36.25 μg) and tetrachloroethylene (40.5 μg) was added. Equilibration took one hour at 110 °C. The respective peak area values were 882,610 (C_2HCl_3) and 2,393,235 (C_2Cl_4). Using eq. 5.7 but now written for amounts (W) instead of concentrations, we find

$$f_i = \frac{W_i}{W_{st}} \cdot \frac{A_{st}}{A_i} = \frac{36.25}{40.5} \cdot \frac{2,393,235}{882,610} = 2.427$$

Sample analysis. The capsule sample (1.31 g) was placed together with 1 mL of water into the headspace vial and 2 μL of a 0.1% tetrachloroethylene solution in cyclohexane (3.24 μg) was added. Equilibration was as above. Figure 5-3 shows the obtained chromatogram; for the GC conditions see the figure caption. The following peak area values were obtained: C_2HCl_3, 64,312 counts; C_2Cl_4, 307,646 counts. The amount of trichloroethylene present is calculated using eq. 5.8, again considering

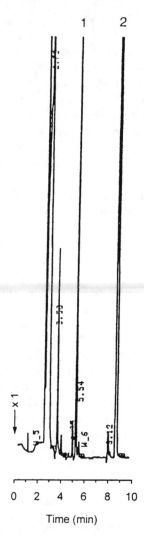

Figure 5-3. Determination of residual trichloroethylene in a gelatin capsule, using the internal standard method, with tetrachloroethylene as the internal standard. For details, see Example 5.2. *HS conditions*: Sample: 1.31 g capsule + 1 mL water + 2 μL of a 0.1% tetrachloroethylene solution in cyclohexane. Equilibration at 110 °C for 1 hour. *GC conditions*: 50 m × 0.32 mm I.D. fused-silica, open-tubular column coated with bonded phenyl (5%) methyl silicone stationary phase; film thickness: 1 μm. Column temperature: 80 °C. Electron-capture detector. *Peaks*: *1* = trichloroethylene; *2* = tetrachloroethylene.

amounts instead of concentrations:

$$W_i = W_{st} f_i \cdot \frac{A_i}{A_{st}} = 3.24 \times 2.427 \cdot \frac{64,312}{307,646} = 1.65 \, \mu g$$

This corresponds to a trichloroethylene concentration of 1.26 ppm in the gelatin capsule.

Blood Alcohol Determination.   Probably the best-known and most widely used application of HS-GC is the establishment of the ethanol content of blood, in the testing of automobile drivers charged with DWI (driving while intoxicated). The classical method was developed by G. Machata in 1964 [1] and in fact, it represented the start of quantitative HS-GC. In subsequent years, further refinements of the method have been published [2,3]. Today Machata's approach is the accepted standard method in most countries (see, e.g., ref. 4).

In the *original Machata method*, 0.5 mL blood was mixed with 0.1 mL of a standard solution of *tert*-butanol (2 mg/mL) used as the internal standard, and the mixture was thermostatted at 60 °C for about 20–30 minutes. The calculation was carried out in the usual way as discussed earlier; however, the calibration factor also incorporated additional corrections related to differences in the vapor pressure of alcohol over an aqueous solution (used in establishing the calibration factor) versus over blood, and to the specific weight of blood (1.057 g/mL) because the sample is pipetted and not weighed into the vial.

In the *modified method* used today in most countries, a smaller blood sample — only 0.1–0.5 mL — is used, which is then diluted with an excess of the aqueous internal standard solution (typically 1:5 or 1:10). In this way, the blood matrix is diluted so much that any effect of matrix variation is eliminated.

Besides *tert*-butanol, sometimes *n*-propanol may be used as the internal standard for blood alcohol determination. Which standard is actually used depends mainly on the separation characteristics of the column applied in the gas chromatograph. Figure 5-4 shows a typical chromatogram using *n*-propanol as the internal standard.

5.3 External Standard Method

In the external standard method, the analysis of the unknown sample is compared with the analysis of the standard sample having the same matrix and containing the same analyte, in known concentration. Since peak area and amount (concentration) are proportional, the amount (concentration) of the analyte can be readily calculated from the two peak areas and its amount (concentration) in the standard solution. Since we have the same analyte in

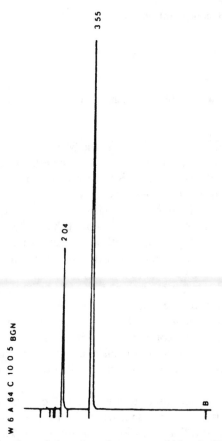

Figure 5-4. Determination of the ethanol content of blood. *HS conditions*: Sample: 0.5 mL blood, diluted 1:5 with an aqueous solution of *n*-propanol (4 mg/mL). Equilibrated at 55 °C for 12 minutes. *GC conditions*: Column: 30 m × 0.53 mm I.D. fused-silica, open-tubular, coated with bonded methyl silicone stationary phase; film thickness: 3 μm. Column temperature: 45 °C. Carrier gas: Helium, flow rate: 7.5 mL/min. Flame-ionization detector. *Peaks*: *1* = ethanol, *2* = *n*-propanol.

Source: Reproduced with permission of D. J. Brown, W. C. Long, and *Journal of Analytical Toxicology* [4].

both samples, the calibration factor will be the same. Therefore:

$$\frac{C_{i(ex)}}{A_{i(ex)}} = \frac{C_{i(o)}}{A_{i(o)}} \tag{5.11}$$

$$C_{i(o)} = C_{i(ex)} \cdot \frac{A_{i(o)}}{A_{i(ex)}} \tag{5.12}$$

In these equations, subscript (*ex*) refers to the external standard while subscript (*o*) refers to the actual sample. Naturally, both measurements must be carried out under identical conditions.

The use of the external standard method has a number of advantages. For example, it works even if the chromatogram of the sample is "too crowded" and one cannot find a place for the internal standard peak or, if for any reason, one does not want to add a new compound (the standard) to the sample. It is particularly convenient when a number of samples with the same qualitative but different quantitative composition are to be analyzed: in this case, one needs only a single external standard for the whole series of determinations.

The main difficulty associated with the external standard method is the need to reproduce the matrix of the sample. This problem was discussed in Section 4.5, and we would like to draw the reader's attention to that discussion once again.

Four examples are shown here to illustrate various questions related to the external standard method; since the calculation of the sample concentration is straightforward, it is omitted.

Figure 5-5 demonstrates a case of a neat matrix that is readily available. Fresh engine oil does not contain any volatile aromatic hydrocarbons and

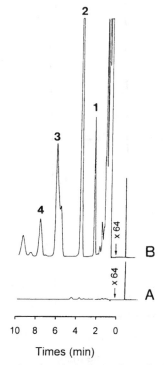

Figure 5-5. Determination of aromatic hydrocarbons in engine oil. (A) fresh engine oil containing no aromatic compounds; (B) used engine oil. *HS conditions*: 1 mL sample, equilibrated at 60 °C for 30 minutes. *GC conditions*: 2 m × 1/8 in. O.D. packed column containing 15% poly(ethylene glycol) on Celite 60/80 mesh. Column temperature: 75 °C. Flame-ionization detector. *Peaks*: *1* = benzene; *2* = toluene (0.1 wt%), *3* = *m/p*-xylene; *4* = *o*-xylene.

thus, it can be used as the solvent when preparing a standard solution of these compounds. In this way, both the external standard and the actual sample will have the same matrix: thus, slight variations in the original composition of the oils can be ignored.

Figure 5-6 is another example of a neat matrix being available: it illustrates a very important application of HS-GC in the 1980s, the determination of traces (0.9 ppm) of tetrachloroethylene in a Spanish olive oil. Again pure olive

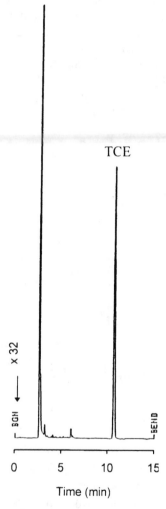

Figure 5-6. Determination of 0.9 ppm residual tetrachloroethylene (TCE) in olive oil, using the external standard method. *HS conditions*: 5 mL sample, equilibrated at 80 °C for 30 minutes. *GC conditions*: 50 m × 0.32 mm I.D. fused-silica, open-tubular column coated with bonded methyl silicone stationary phase; film thickness: 1 μm. Column temperature: 70 °C. Electron-capture detector.

oil, without any residual solvent, could be easily found and thus used to prepare the external standard.

In the third example, the matrix effect was eliminated by diluting the sample. An electrolytic (anodic) plating bath contains a number of nonvolatile sample constituents such as salts, detergents, and pigments that could influence the distribution of the volatile organic compounds present. The first attempt to apply the FET failed (see Section 6.4) due to residual adsorption effects. However, the concentration of the volatile compounds of interest is high enough to permit a 1:10 dilution with water and by this, any matrix effect is eliminated. Figure 5-7 shows the obtained chromatogram; since the detector attenuation was $\times 32$, even more dilution would be feasible. Because now the matrix effect could be ignored, a simple aqueous solution of the analytes could be used as the external standard.

Finally, the last example (Figure 5-8) demonstrates the determination of methanol in adulterated wine. Wine is a diluted aqueous solution, and the only main components that would influence the partition coefficient of methanol are water and ethanol; therefore, a 10% aqueous ethanolic solution containing 0.2% (wt/vol) methanol was used as the external standard. Since we were interested only in the methanol peak, the column was backflushed after its elution to speed up the analysis.

5.4 Standard Addition Method

In the standard addition method, the analysis of the original sample is followed by the analysis of the same sample to which known amounts of the analyte are added; all measurements are carried out under identical conditions.

The standard addition method is a universal procedure in headspace measurement and has been recommended since the early days of quantitative headspace analysis [5]. The determination is carried out in an identical matrix, and thus no response (calibration) factors are needed for the calculation. Peak area and amount of the analyte are proportional: from this relationship one can then directly calculate the original amount present.

5.4.1 Single Addition

First we investigate what happens when the analysis of the sample is compared to the analysis of the same sample to which a known amount of the analyte was added. Let us denote the original amount present as W_o and the corresponding peak area as A_o. We added W_a amount of the analyte to the original sample, obtaining now a peak area $A_{(o+a)}$. Based on the proportionality between analyte amount and peak area, we can write:

$$\frac{W_o}{A_o} = \frac{W_o + W_a}{A_{(o+a)}} \tag{5.13}$$

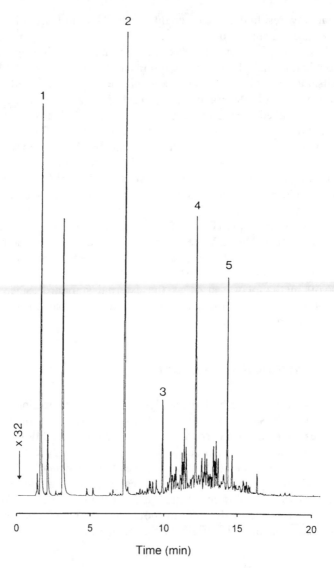

Figure 5-7. Determination of the volatile organic constituents present in an anodic electrolytic plating bath. *HS conditions*: Original sample diluted 1:10 with water; 2 mL of the diluted solution analyzed. Equilibration at 90 °C for 60 minutes. *GC conditions*: 25 m × 0.25 mm I.D. fused-silica, open-tubular column coated with bonded methyl silicone stationary phase; film thickness: 1 μm. Column temperature: 2 minutes isothermal at 50 °C, then programmed at 8 °C/min to 180 °C. Split sampling. Flame-ionization detector. *Peaks* (concentration in the original sample): *1* = isopropyl alcohol (0.03%), *2* = butyl cellosolve (0.33%), *3* = n-decane, *4* = n-undecane, *5* = n-dodecane. Total amount of "hydrocarbon oil" (the peaks after peak 2) is 0.2%.

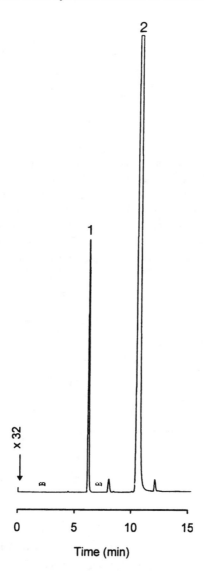

Figure 5-8. Determination of methanol in adulterated wine, using the external standard method. *HS conditions*: 5 mL sample, equilibrated at 60 °C for 30 minutes. GC conditions: Two 25 m × 0.32 mm I.D. fused-silica, open-tubular columns coated with bonded methyl silicone stationary phase; film thickness: 1 μm. Backflush configuration. Column temperature: 60 °C. Flame-ionization detector. *Peaks*: *1* = methanol (0.2% wt vol); *2* = ethanol.

Thus:

$$W_o = W_a \frac{A_o}{A_{(o+a)} - A_o} = W_a \cdot \frac{A_o}{\Delta A} \tag{5.14}$$

where

$$\Delta A = A_{(o+a)} - A_{(o)} \tag{5.15}$$

If the amount W_o of the analyte is added as a solution to the second sample, then we must also add the same volume of the solvent (naturally, without the analyte) to the first (original) sample, because the addition of this volume of solvent will change both the phase ratio and the matrix.

The calculation may also be carried out relative to an internal standard (we use here the term *normalization standard*, and the symbol R), the concentration of which remains constant in both samples: in this case the peak area ratios can be used instead of the absolute peak area:

$$R_o = A_o / A_{st} \tag{5.16}$$

$$R_{(o+a)} = A_{(o+a)} / A_{st} \tag{5.17}$$

Substituting these relationships into eq. 5.13 we obtain:

$$\frac{W_o}{R_o} = \frac{W_o + W_a}{R_{(o+a)}} \tag{5.18}$$

and

$$W_o = W_a \cdot \frac{R_0}{R_{(o+a)} - R_o} \tag{5.19}$$

The normalization standard may be a sample component, or it may be added as a solution to the sample. In this case, it is again important to also add the same volume of the solvent (without the standard) to the first sample, because even a small change in the sample volume will change both the phase ratio and the matrix. However, any peak (other than that of the analyte) present in the chromatogram may also be used as the normalization standard; naturally, in this case, no solvent is to be added to the sample.

Figure 5-9 shows the analysis of a massage cream to determine the amount of camphor present. Here, one of the peaks in the chromatogram (eluting at 9 min) was used as the normalization standard (NS).

The use of a normalization standard representing one of the components present is also helpful if the sample is not homogenous. This will be illustrated later, in Chapter 6 (Section 6.3) in connection with the development of methods for the analysis of residual solvents in a printed, laminated plastic film.

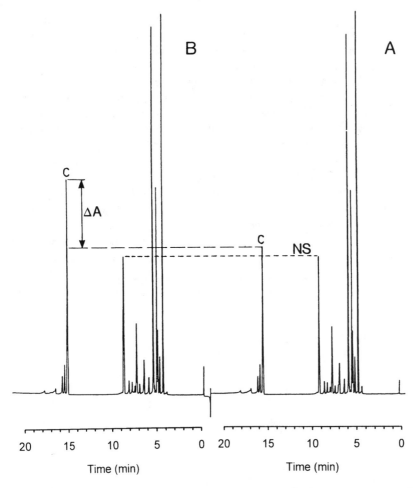

Figure 5-9. Determination of the camphor content of a massage cream, using the standard addition method: $A = 1.0\,g$ of the original sample; $B = 1.0\,g$ of the sample to which 5 mg of camphor was added. *HS conditions*: 1.0 g sample, equilibrated at 80 °C for 1 hour. *GC conditions*: 20 m × 0.25 mm I.D. glass, open-tubular column coated with poly(ethylene glycol) stationary phase; film thickness: 0.2 μm. Column temperature programmed from 70 °C to 150 °C at 5 °C/min. Flame-ionization detector. C = camphor, NS = normalization standard, ΔA = increase in peak area due to the addition of 4.0 mg camphor.

5.4.2 Handling of the Added Standard (GPA and SPA)

A particular advantage of the standard addition method is that the additional amount of the analyte can be added into either the sample phase or the gas phase of the vial. This is so because equilibration is a two-way process: the analyte molecules move from the sample proper into the headspace in contact with it; but there is also a parallel movement from the headspace into the sample proper, and at equilibrium the two movements compensate each other.

To understand this process, let us consider a *Gedankenexperiment*. Into a number of headspace vials we add the same volume of the pure matrix, and then we add the same amount of the analyte *into the matrix* in each vial.* Subsequently, we thermostat the individual vials for increasingly longer times (e.g., using the progressive working mode) and sample the headspace of the individual vials at the end of the thermostatting time. As discussed in Section 4.1, at the start there will be no analyte in the headspace but, with increasing thermostatting time, more and more will diffuse into it until equilibrium is reached. From then on the amount (concentration) of the analyte in the headspace (and also in the sample phase) will remain constant, regardless of the thermostatting time. This situation is illustrated in the lower plot of Figure 5-10 marked SPA (*sample phase addition*).

Next, we repeat the experimental series but now add the same amount of the analyte *into the headspace* of the vials. Thus, although at the beginning no

---

* We assume that the volume of the added analyte is negligible relative to the volume of the original matrix.

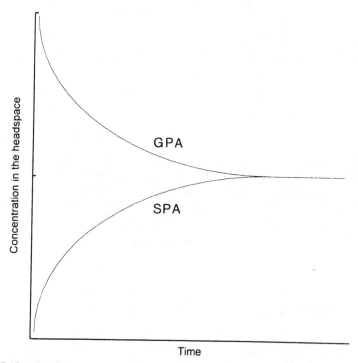

Figure 5-10. Equilibration when adding an aliquot of an analyte to the vial. *SPA* = sample phase addition; *GPA* = gas phase (headspace) addition.

analyte is in the matrix (sample phase), it will soon start to diffuse in from the headspace; as a consequence, the analyte's concentration in the headspace will gradually diminish until equilibrium has been reached. From then on the analyte's concentration in either phase will remain constant regardless of the thermostatting time. This situation is illustrated in the upper plot in Figure 5-10, marked GPA (*gas phase addition*): as expected, it is the mirror image of the SPA plot. At equilibrium the two plots converge.

Naturally, the same process will take place if we add an aliquot of the analyte to the vial containing a sample already containing a certain amount of the analyte: it is irrelevant where the aliquot was placed, into the sample or into the gas phase and, at equilibrium, the same distribution will be achieved.

In general, we use the SPA technique in the case of liquid samples where the added compound is usually dissolved homogeneously in the liquid sample phase. The GPA technique is particularly useful for solid samples where mechanical mixing of, say, a few ppm of a monomer into a solid polymer is practically impossible. Naturally the solid sample must represent a partition system and this needs to be confirmed (e.g., by the MHE technique: see Section 5.5). As discussed in Section 4.1.3, polymers above the glass transition point usually fulfill this requirement.

The next example dealing with the determination of residual vinyl chloride monomer (VCM) in a PVC resin uses both the SPA and GPA approaches to illustrate their equivalency.

Example 5.3

In the first experiment 1 g of the PVC resin was dissolved in the headspace vial in 4 mL of dimethylacetamide (DMA) (vial 1), while in the second experiment, 4 g of dry PVC resin was placed into a headspace vial (vial 2), without any solvent. The solution was equilibrated at 80 °C for 120 minutes and the solid polymer at 110 °C for 60 minutes and the headspace of the vials analyzed in the usual way; the accompanying table gives the obtained VCM peak area values under A_o.

Next, two new vials were prepared by adding the solution of 1 g PVC resin in 4 mL of DMA into vial 3 and 4 g of dry PVC resin into vial 4. However, the DMA solution in vial 3 was spiked with 5 μL of a solution of VCM in DMA, while 5 μL of this solution was added to the headspace of vial 4. The concentration of this VCM solution was 2.083 μg/μL: thus the 5 μL contained 10.41 μg VCM. These samples were again thermostatted and analyzed in the same way as the first two samples: the obtained peak areas are given below under $A_{(o+a)}$. The values in the table represent the mean of three determinations, the numbers in parentheses refer to the serial number of the vial as given above.

	Pure PVC sample, A_o	PVC sample with added 10.41 μg VCM, $A_{(o+a)}$	ΔA
1 g PVC resin + 4 mL DMA	1,994 (*1*)	8,550 (*3*)	6,556
4 g dry PVC resin	20,405 (*2*)	36,442 (*4*)	16,037

For the calculation, we use eq. 5.14:

$$\text{SPA: } W_o = 10.41 \cdot \frac{1{,}994}{6{,}556} = 317 \, \mu g \text{ in 1 g PVC resin } (= 3.17 \, \text{ppm})$$

$$\text{GPA: } W_o = 10.41 \cdot \frac{20{,}405}{16{,}037} = 13.245 \, \mu g \text{ in 4 g PVC resin } (= 3.31 \, \text{ppm})$$

The mean of the two determinations is 3.24 ppm ($\pm 2.16\%$).

A comparison of the peak areas obtained in the two approaches shows the much lower sensitivity with the "solution approach," where the solid sample was dissolved in an excess of solvent. This is, of course, to be expected. However, the results of the approaches agree well and show that with the GPA technique, both high sensitivity and good accuracy were obtained.

Separate investigations have shown that using the GPA approach, the detection limit of VCM in a solid PVC resin (using a 4 g sample) is about 1 ppb with a flame-ionization detector (at highest sensitivity). Such an example is shown in Figure 5-11.

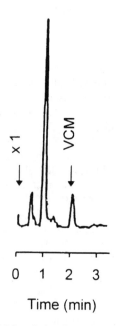

Figure 5-11. Determination of 2.5 ppb ($\mu g/kg$) vinyl chloride monomer (VCM) in a PVC resin. *HS conditions*: Sample: 4 g PVC resin, equilibrated at 110 °C for 60 minutes. *GC conditions*: Two 1 m × 1/8 in. O.D. packed columns in backflush configuration, containing 0.19% picric acid on Carbopak C, 80/100 mesh. Column temperature: 40 °C. Flame-ionization detector.

5.4.3 Determination with Multiple Additions

Using only two measurements — the original sample and one with the addition of a known amount of the analyte — deviation in sampling or linearity will directly influence the result. On the other hand, if increasing amounts of the analyte are added to the original sample and the results are evaluated by linear regression analysis, individual deviations are compensated. Therefore, we recommend the use of the single addition procedure only in routine analyses after the linear range of the analyte over the concentration range of interest has been confirmed by multiple measurements and their linear regression analysis.

To explain the evaluation of multiple measurements with increasing amounts of additions, we start with eq. 5.13:

$$\frac{W_o}{A_o} = \frac{W_o + W_a}{A_{(o+a)}} \tag{5.13}$$

where subscripts o and a refer to the original sample and the added amount of the analyte. Reorganizing eq. 5.13 gives:

$$A_{(o+a)} = \frac{A_o}{W_o} W_a + A_o \tag{5.20}$$

which corresponds to a linear equation of the $y = ax + b$ type, where:

$x = W_a$

$y = A_{(o+a)}$

$a = A_o/W_o$

$b = A_o$

Carrying out a number of additions, the data (W_a vs. $A_{(o+a)}$) can be evaluated by linear regression analysis; then from the slope a and intercept b of the linear plot value of W_o, the amount of analyte in the original sample, can be calculated:

$$W_o = b/a \tag{5.21}$$

The value of W_o may also be established graphically. If the values of W_a are plotted against $A_{(o+a)}$, the y intercept will be A_o and the intercept with the abscissa will be equal to W_o. This can be derived by substituting $A_{(o+a)} = 0$ into eq. 5.20. In this case

$$W_a \cdot \frac{A_o}{W_o} = -A_o \tag{5.22}$$

and

$$W_o = -W_a \tag{5.23}$$

The value is negative because it is on the negative side of the abscissa (see Figure 5-13).

In the example that follows, the aim was to determine residual solvents (acetone and dichloromethane) in a steroid drug. The solution approach was selected to handle the sample: acetic acid was found to give the best solubility for this particular steroid sample, which was difficult to dissolve in other solvents. To compensate for the diminished sensitivity and for the effect of dissolution, a rather concentrated solution (10%) was prepared; standard addition was selected to eliminate any possible matrix effect of this fairly concentrated steroid solution [6].

Example 5.4

In a number of separate vials, 200 mg of the steroid drug was dissolved in 2.0 mL of acetic acid and 0, 5, 10, 15 and 20 µL of the standard solution of the analyte(s) with 1 vol% concentration were added. The vials were equilibrated at 80 °C for 30 minutes; for the conditions of the GC analysis see the caption of Figure 5-12.

These data are for dichloromethane only: since the density of CH_2Cl_2 is 1.3348 g/mL, 1 µL of a 1 vol% solution will contain 0.01 µL = 0.01335 mg

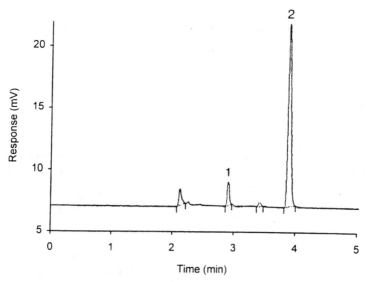

Figure 5-12. Determination of the residual solvent content in a steroid drug, dissolved in acetic acid, by multiple standard addition, as described in Example 5.4. Chromatogram of the original sample without addition. *HS conditions*: Sample: Solution of 200 mg drug in 2.0 mL acetic acid. Equilibration at 80 °C for 30 minutes. *GC conditions*: 50 m × 0.32 mm I.D. fused-silica, open-tubular column coated with bonded poly(ethylene glycol) stationary phase; film thickness: 1 µm. Column temperature: 80 °C. Flame-ionization detector. *Peaks* (concentration): *1* = acetone (43 µg/g); *2* = dichloromethane (473 µg/g).

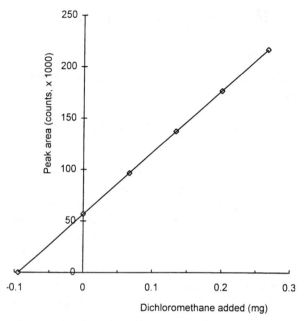

Figure 5-13. Linear regression plot for the determination of dichloromethane in a steroid drug, by multiple standard addition.

dichloromethane. Table 5-4 lists the results; the regression plot is given in Figure 5-13. The calculated value of the amount of CH_2Cl_2 is

$$W_o = b/a = 0.09457 \text{ mg}$$

corresponding to a sample concentration of 472.9 μg/g. A similar measurement gave an acetone concentration of 43 μg/g.

Table 5-4   Analysis of a steroid sample for
              dichloromethane, using multiple
              standard addition (Example 5.4)*

Amount of CH_2Cl_2 added (mg)	Peak area (counts)
0	56,919
0.067	96,495
0.134	137,007
0.200	176,366
0.267	216,819
Linear regression	
Correlation coefficient	0.999993
Slope a	599,206
Intercept b	56,667

* For the linear regression plot see Figure 5-13.

5.5 Multiple Headspace Extraction (MHE)

5.5.1 Principles of MHE

We have already discussed the principles and theoretical background of multiple headspace extraction (see Section 2.6). To summarize: it is dynamic gas extraction, carried out stepwise. After a number of consecutive measurements have been performed, the peak area corresponding to the total amount of the analyte present is established. In this way, any effect of the matrix is eliminated and the calculation of the total amount of the analyte present depends only on a calibration (response) factor, just as in any quantitative measurement by GC. Now we repeat the most important basic relationships of MHE, already discussed in Chapter 2. As derived there, quantitative evaluation of the measurement series is based on the linear regression analysis of the peak area obtained in the consecutive measurements.

$$\ln A_i = -q \cdot (i - 1) + \ln A_1 \tag{5.24}$$

where A_1 is the peak area obtained in the first measurement and q is the slope of the $\ln A_i$ versus $(i - 1)$ plot. From the value of the slope we obtain the quotient Q:

$$-q = \ln Q \tag{5.25}$$

$$Q = e^{-q} \tag{5.26}$$

which is equal to the ratio of the peak area obtained in two consecutive measurements:

$$Q = \frac{A_2}{A_1} = \frac{A_3}{A_2} = \frac{A_{(i+1)}}{A_i} = e^{-q} \tag{5.27}$$

In this calculation the quotient Q is established from the linear regression analysis, compensating for random variations in the values of two consecutive peak areas (eq. 5.27). From all the peaks in the series of an MHE analysis, it is the first peak A_1 that is particularly prone to experimental errors. This is why we carry out a multipoint measurement. By applying linear regression calculation, we obtain the intercept A_1^* and use this value rather than the experimental first area value A_1, which includes the statistical random variation and should therefore not be used as the starting value for further extrapolation, regardless how small the difference between the values may be. From the value of Q and the intercept A_1^*, the area total corresponding to the total amount of analyte present can be calculated:

$$\sum_{i=1}^{i=\infty} A_i = \frac{A_1^*}{1 - Q} = \frac{A_1^*}{1 - e^{-q}} \tag{5.28}$$

If high accuracy is not required, or if it has been established that the linearity of eq. 5.24 is sufficiently good, the area total may be calculated from the results of only two consecutive measurements. In this, *two-point measurement*, the area total is calculated as:

$$\sum_{i=1}^{i=\infty} A_i = \frac{A_1^2}{A_1 - A_2}$$
(5.29)

Naturally we use in this case, the experimental peak areas (A_1) and (A_2).

In the subsections that follow, we give many practical examples for the application of MHE including all the basic data (such as the peak area values) in the pertinent tables. This is to enable interested readers to use these data for their own calculations. It might be of particular interest to compare the result of a more accurate multipoint MHE analysis with that of the simple two-point analysis.

5.5.2 Calibration in MHE

In MHE we obtain the total peak area, which is directly proportional to the total amount of the analyte present in the sample. The actual amount can be established by the proper calibration. There are three ways to do this.

5.5.2.1 External Standard

In the first case an *external standard* is used, which is carried through an MHE process similar to that for the sample proper. However, because an exhaustive gas extraction is performed with both samples, this standard does not have to contain the same matrix as the sample: for example, a simple vapor standard prepared in the vial using the total vaporization technique (TVT: see Section 4.6.1) can be used. MHE measurements with this standard sample will give the sum of the peak areas (ΣA_{ex}) corresponding to the amount of analyte present in the standard (W_{ex}). Because peak area and amount of the analyte are proportional, the amount of the analyte in the sample (W_i) can be calculated from these data and the sum of the peak areas obtained for the sample (ΣA_i):

$$\frac{W_i}{\Sigma A_i} = \frac{W_{ex}}{\Sigma A_{ex}}$$

$$W_i = \frac{\Sigma A_i}{\Sigma A_{ex}} \cdot W_{ex}$$
(5.30)

Finally, the analyte's concentration in the sample can be calculated as

$$C_i\% = 100 \frac{W_i}{W_{sample}} = 100 \frac{\Sigma A_i}{\Sigma A_{ex}} \cdot \frac{W_{ex}}{W_{sample}}$$
(5.31)

Note that in eq. 5.31, W_{sample} is not the amount of the analyte in the sample phase (for which we generally use the symbol W_s), but the amount of the *total sample*.

The following example demonstrates the MHE method with external standard calibration for the determination of residual ethylene oxide (EO) in surgical material, in this case, a PVC tube. EO is used in the sterilization of such items. In this determination an EO standard solution, prepared as described in Section 4.4.2, was used as the external vapor standard. Correction for the sample volume was carried out as described shortly (Section 5.5.3).

Example 5.5

A 1.0 g piece of PVC tube (0.70 mL) was cut into small pieces and transferred into the headspace vial. A 8 μL aliquot of an aqueous EO solution having a concentration of 1.03 mg/mL was used as the external vapor standard: it contained 8.24 μg of EO. Equilibration of both the sample and the standard solution was carried out at 80 °C for 90 minutes.

The analytical results are listed in Table 5-5 together with the linear regression data; the corresponding plots are shown in Figure 5-14.

The following calculation of the sum of peak areas makes use of the intercept value A_1^*, which is obtained from the linear regression calculation (Table 5-5):

sample:
$$\Sigma A_i = \frac{146{,}103}{1 - 0.4347} = 258{,}452$$

calibration standard:
$$\Sigma A_{ex} = \frac{72{,}636}{1 - 0.3025} = 104{,}138$$

The value of ΣA_{ex} must be corrected for the sample volume of 0.7 mL (see Section 5.5.3: eqs. 5.35 through 5.38) considering a vial with a volume of 22.3 mL and thus

Table 5-5   MHE measurement of residual ethylene oxide in a surgical PVC tube (Example 5.5)*

		Peak area (counts) for:	
			External vapor calibration
i		Sample	standard
1		151,909	75,061
2		63,127	21,100
3		26,802	6,583
4		10,963	2,027
5		5,768	613
6		2,240	
Linear regression			
Correlation coefficient r		−0.999142	−0.999897
Slope q		−0.83304	−1.19581
$Q = e^{-q}$		0.4347	0.3025
Intercept A_1^*		146,103	72,636

* For the linear regression plots, see Figure 5-14.

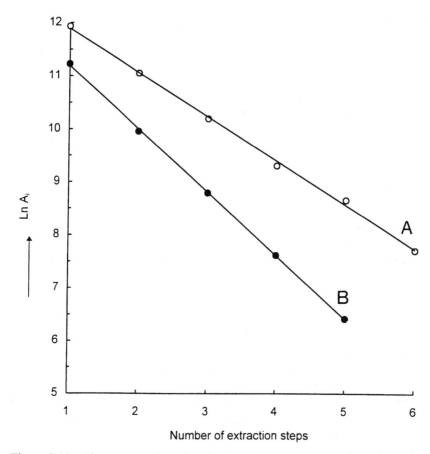

Figure 5-14. Linear regression plots for MHE measurements of residual ethylene oxide (EO) in a surgical PVC tube, using an EO solution as the external standard (Example 5.5). A = PVC tube, B = external vapor standard (EO).

the corrected area total ΣA_{ex}^x is obtained

$$f_V = 1.0324$$

$$\Sigma A_{ex}^x = 1.0324 \times 104{,}138 = 107{,}512$$

In other words, the peak area corresponding to 8.24 μg of EO is 107,512 counts; accordingly, the amount of EO in the sample is (eq. 5.30):

$$\frac{258{,}452}{107{,}512} \cdot 8.24 = 19.81 \ \mu g/g$$

In practice it is not necessary to carry out six or even nine consecutive determinations (unless the aim is to check whether the HS system behaves linearly): three or four measurements usually is enough. Linear regression analysis of only the first three measurements of Example 5.5 gives the following results:

Example 5.5′

Sample	
correlation coefficient r	-0.99997
intercept A_1^*	151,367
slope q	-0.86741
$Q = e^{-q}$	0.4200
ΣA_i	260,978

Calibration standard	
correlation coefficient r	-0.99969
intercept A_1^*	75,768
slope q	-1.21689
$Q = e^{-q}$	0.2961
ΣA_{ex}	107,640
ΣA_{ex}^x	111,128

Total amount of EO in the sample: 8.24 g
Concentration of EO in the 1.0 g sample: $19.35\ \mu g/g$.
The deviation from the six-point determination ($19.81\ \mu g/g$) is only 2.3%.
The simplified, two-point calculation gives the following results (naturally, now using the measured values of A_1.

$$\Sigma A_i = 259{,}921$$

$$\Sigma A_{ex} = 104{,}412$$

$$\Sigma_{ex}^x = 107{,}794$$

EO concentration in the 1 g sample: $19.87\ \mu g/g$.

The concentration of EO in the PVC tube calculated in this way is identical with the result of the multipoint calculation. As mentioned earlier, the agreement between the three calculations depends very much on the deviation of the individual measurements from the corresponding values of the regression plot and therefore, the two-point calculation should be used only in routine analysis, after checking the closeness of this approach to the result of multi-point measurements. The plots shown in Figure 5-14 represent an ideal partition system (corresponding to case D in Figure 5-15: see Section 5.5.6). This means that in either case, practically any quantitative technique can be used for routine analysis: one may use MHE, with three- or two-point measurement, but one may also use the internal standard or standard addition methods. For example, the latter method with three additions gave an EO concentration of $19.95\ \mu g/g$.

5.5.2.2 *Internal Standard*

The second possibility for the quantitative evaluation of MHE measurements is the use of an *internal standard* added to the sample. In this case, both the analyte and the standard will undergo multiple gas extraction, and the calculation of the amount of the analyte is carried out similarly to the way it is done in the regular internal standard method (cf. eq. 5.8) except that now the sum of the respective peak areas is used:

$$W_i = W_{st} \cdot f_i \cdot \frac{\Sigma A_i}{\Sigma A_{st}} \tag{5.32}$$

The *response factor** (f_i or RF) can be established by preparing a vapor standard using the total vaporization technique, with known amounts of the standard and analyte present, and carrying it through an MHE procedure. The response factor can be calculated as follows:

$$RF_i = \frac{W_i^c}{W_{st}^c} \cdot \frac{\Sigma A_{st}^c}{\Sigma A_i^c} \tag{5.33}$$

where superscript c indicates that these values refer to the separate calibration measurement.

A particular advantage of this method of operation is that one obtains the response factor at a number of different concentration levels. Since in a linear HS system the value of the response factor should be independent of concentration, this can be checked by comparing the actual A_{st}^c/A_i^c ratios at each extraction step: it should be constant. For an example, see Section 6.2.

5.5.2.3 *Standard Addition*

Finally, the third possibility is to use *standard addition*. Here we do not need a calibration factor: the amount (concentration) of the analyte present can be directly established as discussed in Section 5.4 (cf. eq. 5.14), except that now the sum of the peak areas is used:

$$W_o = W_a \cdot \frac{\Sigma A_o}{\Sigma A_{(o+a)} - \Sigma A_o} \tag{5.34}$$

where W_o is the amount of analyte present in the original sample and W_a is the amount of analyte added. We have to carry out two sets of MHE measurements: the first with the original sample and the second with the same amount of sample to which W_a amount of the analyte was added; the respective sums of the peak areas are ΣA_o and $\Sigma A_{(o+a)}$. If a solid sample is analyzed, then the gas phase addition (GPA) method is the preferable way to add the additional amount of analyte to the sample.

---

* We now call it a response factor because here it simply reflects the difference in the detector response for the different compounds.

This third possibility of using the standard addition in MHE is of more theoretical than practical use for quantitative analysis, because each of these techniques is applied alternatively rather than in combination. Both techniques have their own pros and cons and the following additional arguments concerning sample handling problems and sample throughput may decide which is preferable for a particular sample.

Sample Handling

If reproducible sample preparation is a problem (e.g., with viscous or inhomogenous samples), the MHE technique that is carried out from a single sample may be preferable. Standard addition requires the reproducible preparation of several sample aliquots and as a consequence the reproducibility of sample handling, which depends also on the skill of the operator, is included in the analytical result. The values for the precision (e.g., the relative standard deviations), therefore, are in general better for the MHE results, since the analyical result is derived from a single sample only and thus includes only the instrumental precision.

Sample Throughput for Automatic Routine Analysis

Standard addition needs several vials — at least two- for each sample, and this reduces the number of places in an automated headspace sampler. If such an instrument can accept, for example, 30 vials, then in fact only 15 samples can be processed automatically. If linear regression technique is applied with say, only two additions, there must be a total of three vials including the pure sample for each sample, and the sample throughput is reduced to 10 samples. If this is not acceptable, the MHE alternative may be advantageous, provided that the time expenditure is not the limiting factor. The series of consecutive analyses in MHE is carried out from the same vial, but this procedure does not allow the time-saving overlapping working mode (see Section 3.4.2), in contrast to the standard addition technique, where separate vials are used for each determination. Therefore samples with an excessively long equilibration time (e.g., solid samples), may cause an excessively long total analysis time with MHE. If, on the other hand, the simplified *two-point measurement* can be applied, this drawback of the MHE technique becomes less important.

We can see from this short discussion that in practice many arguments should be taken into account before deciding on the final analytical technique to solve a particular analytical problem. Therefore, a systematic strategy is desirable, and MHE with standard addition may be more useful for preliminary investigations rather than for practical quantitative analysis. This application, however, is illustrated later, in Chapter 6 (Examples 6.2 and 6.3), when we discuss method development.

5.5.3 The Use of Gaseous External Standards in MHE

Since in an MHE determination the matrix effects play no role, the calibration standards can be simplified to a single compound, the analyte. Such a calibration standard can easily be prepared by the total vaporization technique (see Section 4.6.1), injecting a few microliters of the pure analyte, or its solution, into an empty headspace vial and assuring by the proper selection of the thermostatting temperature that it is completely evaporated.

We have used this method already (Example 5.5): an 8-μL aliquot of an aqueous EO solution served as the calibration standard; at the thermostatting temperature of 80 °C this small volume of the solution will completely evaporate.

Correction for Sample Volume. If such a vapor standard is used as the external calibration standard, correction may be needed for the actual sample volume, unless it is very small. The reason for this can be explained in the following way.

Let us assume that the total amounts of the analyte in the sample and the standard are W_i and W_{ex}, and the respective total peak areas obtained are ΣA_i and ΣA_{ex}. In the vial containing the sample, the analyte vapor is distributed in its headspace volume ($V_G = V_V - V_S$, where V_V and V_S are the respective volumes of the vial and the sample), while in the vial containing the standard, its vapor is distributed in the volume of the vial (V_V). Thus the corresponding gas phase concentrations of $C_{G,i}$ and C_{Gex} can be written based on the fundamental laws of headspace analysis (cf. eqs. 2.19 and 2.17) as:

$$\Sigma A_i \propto C_{G,i} = \frac{W_i}{V_V - V_S} \tag{5.35}$$

$$\Sigma A_{ex} \propto C_{G,ex} = \frac{W_{ex}}{V_V} \tag{5.36}$$

Let us now assume that $W_i = W_{ex}$. It is obvious from eqs. 5.35 and 5.36 that we would obtain a smaller concentration, hence a smaller peak area for the vial containing the vapor standard, because $V_V > (V_V - V_S)$. Thus, if the volume of the sample cannot be neglected relative to the vial's volume, some correction is necessary to compensate for this effect. There are two possibilities.

In the first, we simulate the sample volume, by adding the appropriate volume of an inert material (e.g., glass beads) to the vial containing the vapor standard. If, however, these beads have a rough surface, adsorption effects should not be ignored.

The second possibility is to calculate a corrected peak area total for the standard (ΣA_{ex}^x), multiplying the established ΣA_{ex} value by a factor correcting for volume differences:

$$f_V = \frac{V_V}{V_V - V_S} \tag{5.37}$$

$$\Sigma A_{ex}^x = f_V \cdot \sum A_{ex} \tag{5.38}$$

An example for this usage was given in Example 5.5, representing the determination of residual ethylene oxide in a surgical PVC tube.

Such a volume correction, however, is unnecessary if an external vapor standard is prepared by flushing the vial with a gas standard of known composition. In this case the concentration of the analyte is already known rather than being prepared by evaporation of a certain amount of the analyte W_{ex} in the vial volume V_V.

5.5.4 The Role of Quotient Q

The geometric progressions used in the calculation of the sums of the peak areas in MHE are characterized by the quotient Q representing the ratio of two subsequent peak areas:

$$Q = \frac{A_2}{A_1} = \frac{A_3}{A_2} = \frac{A_{(i+1)}}{A_i} \tag{5.27}$$

We have also seen the direct relationship between q, the slope of the MHE plot, and quotient Q:

$$-q = \ln Q \tag{5.25}$$

$$Q = e^{-q} \tag{5.26}$$

The key problem in headspace analysis is whether we have a linear system. In this respect "linear" means that in the concentration range represented by the sample in the vial, the partition coefficient is independent of the analyte concentration. In this case the value of the area ratio Q is constant in the whole range. In addition, the actual value of Q (i.e., the slope of the MHE plot) and the relative position of the two plots (sample and standard) reveal important information with regard to the correctness of the headspace analysis. Therefore, it is important to investigate the role of Q and its relationship to other parameters.

5.5.4.1 Relationship Between Q and Pressures

To understand the influence of Q on the MHE results, we first consider the pressure changes occurring during the measurement. The absolute pressures are used here.

After the transfer of an aliquot of the headspace into the column, the vial is pressurized again and the pressure in it is p_h. The subsequent venting process can be visualized as a gas expansion: the gas present in the headspace volume (V_G) at pressure p_h expands to a volume of V_e at pressure p_o. Volume V_e consists of two parts: the volume of the headspace in the vial (V_G) and the volume that was vented (V_{vent}):

$$V_e = V_G + V_{vent} \tag{5.39}$$

Based on the basic gas laws we can write

$$V_G \cdot p_h = (V_G + V_{vent}) \cdot p_o$$

$$\frac{V_G}{V_G + V_{vent}} = \frac{p_o}{p_h} = \rho \tag{5.40}$$

Thus, ρ is a relative pressure expressing the pressure after venting as a function of the pressure in the vial before venting. Because $p_h > p_o$, the value of ρ is always less than unity. If, for example, we assume $p_h = 200\,kPa$ (absolute pressure) as a typical column inlet pressure and $p_o = 100\,kPa$ (atmospheric pressure), then a typical value for ρ will be 0.5.

The pressure in the pressurized vial should be released fast enough to avoid any instantaneous change in the sample distribution between the two phases in the vial during venting. On the other hand, when the pressure in the vial has approached atmospheric pressure, the connection of the vial to atmosphere should be interrupted again after few seconds (e.g., 5 s) to avoid further expansion of the headspace to atmosphere by the increasing vapor pressure of the sample matrix (e.g., water), which reestablishes again slowly after the headspace was released down to atmospheric pressure by venting.

In eq. 5.27 we expressed Q as the ratio of two consecutive peak areas. Since in HS-GC peak area is proportional to the concentration of the analyte in the headspace, we can also write

$$Q = \frac{C_{G2}}{C_{G1}} = \frac{C_{G3}}{C_{G2}} = \frac{C_{G(i+1)}}{C_{G,i}} \tag{5.41}$$

$$Q = \frac{C_{G(i+1)}}{C_{G,i}} = \frac{A_{(i+1)}}{A_i} \tag{5.42}$$

It can be deduced [7] that the quotient Q can be expressed as a function of the partition (distribution) coefficient K of the analyte, the phase ratio β of the vial, and $\rho = p_o/p_h$ (cf. eq. 5.40), which, within a system, are constant:

$$Q = \frac{(K/\beta) + \rho}{(K/\beta) + 1} \tag{5.43}$$

Since $\rho < 1$, the value of Q is also less than unity and therefore, $q\,(= \ln Q)$ will be negative.

5.5.4.2 Value of Q in the Case of Total Vaporization

An interesting question with respect to Q arises when the total vaporization technique is used. In this case, there is only a single phase in the vial: the whole sample was fully evaporated. Therefore, we have no distribution between two phases, and thus $K = 0$. In such a case, eq. 5.43 can be written as follows:

$$Q = \rho/1.0 = p_o/p_h \tag{5.44}$$

In other words, in the TVT, the quotient Q is equal to the ratio of the vial pressure after venting relative to its pressure before venting. This illustrates the importance of exactly reproducing the condition in MHE, because only in this way can the constancy of Q be assured.

The correct value of Q can be controlled independently from any analytical measurements simply by the ratio of the two pressure values in eq. 5.44, which should be identical with the area ratio Q from such a vapor standard. If Q is higher, the vent time is apparently too short and the pressure in the vial has not yet reached atmospheric pressure, if it is lower, a leak in the system (e.g., a leaking vial closure) may cause an additional loss of the sample vapor through the unwanted opening.

The relationship in eq. 5.44 makes it possible to determine the barometric pressure from the value of Q and the (absolute) vial pressure p_h, which is the sum of the atmospheric pressure and the vial pressure Δp^*:

$$p_h = \Delta p + p_o \tag{5.45}$$

Thus:

$$Q = \frac{p_o}{\Delta p + p_o} \tag{5.46}$$

or

$$p_o = \Delta p \cdot \frac{Q}{1 - Q} \tag{5.47}$$

This means that in the case of a multiple headspace extraction measurement, utilizing the total vaporization technique, the atmospheric pressure can be established from the measured data.

As an example let us consider the measurement outlined in Section 9.7.2. According to Table 9-5 $Q = 0.4442$ and the vial pressure was $\Delta p = 120.0\,\text{kPa}$. Because splitless injection onto a fused-silica, open-tubular column (cf. Figure 3-12II) was applied, the vial pressure is identical to the pressure drop Δp along the column. This gives:

$$p_o = 120 \cdot \frac{0.4442}{1 - 0.4442} = 95.9\,\text{kPa}$$

The actual measured atmospheric pressure was $p_o = 96.7\,\text{kPa}$; in other words, the value established from the MHE measurement was within 0.8% of the measured value. Naturally one would not use a headspace sampler as a barometer, but this application may help to control the instrumental parameters.

---

* Again (cf. Section 3.5.6) Δp is the pressure in the pressurized vial as shown by the pressure readout and is the difference between the absolute vial pressure and the atmospheric pressure.

5.5.4.3 The Relative Position of the MHE Plots as a Function of Q

In MHE we have two geometric progressions, one for the analyte in the sample and the other for the standard, and each geometric progression [corresponding to a linear plot in the system of ln A_i versus $(i - 1)$] is characterized by its own slope q or its Q value. Investigation of eqs. 5.43 and 5.44 permits the drawing of certain conclusions concerning the relative position of the MHE plots. We shall specifically consider here what happens when each of the three possibilities for MHE calibration is used.

In the first case, an *external vapor standard* prepared by total vaporization is used. Depending on the situation in the sample vial, we have two possibilities. If the sample vial represents a two-phase system, with the analyte partitioning between the headspace and the sample phase, then the value of Q for the vapor standard will be smaller than for the sample. This can be deduced from eqs. 5.43 and 5.44. For example, taking $K/\beta = 2.0$ and $\rho = 0.5$ as typical values:

sample:
$$Q_S = \frac{2.0 + 0.5}{2.0 + 1} = 0.83$$

vapor standard: $Q_{st} = 0.5$

and the respective slope values are $q_s = -0.19$ and $q_{st} = -0.69$.

This means that in general, the numerical values of Q for the standard plot will be smaller than for the sample plot. Typical values were shown in Figure 5-14, the situation corresponding to case D in Figure 5-15.

With an increasing value of K/β, the value of Q for the sample plot will approach unity, giving a slope value approaching zero. In other words, the sample plot will be almost parallel to the abscissa. This means that there is only very little difference between two consecutive peak areas, property obviously detrimental to the accuracy of the MHE measurement. This situation corresponds to case C in Figure 5-15.

Another possibility is that the analyte in the *sample vial* is also fully evaporated, using the FET (see Section 4.6.2). In this case, the situation in both vials will be the same: there is no further partitioning in either, and thus $K = 0$. Since both vials are now analyzed under identical conditions, the value of Q will be the same for them and the two plots will be (almost) parallel. This situation corresponds to case E in Figure 5-15.

In the second case, calibration is carried out using an *internal standard* and we deal with a partition system. Here, the matrix is the same; however, the partition coefficient of the internal standard versus the analyte will be different: thus, the slope of the two plots will be different. Such a situation will be illustrated in Figure 6-3, where the concentration of residual styrene monomer in polystyrene is determined, using added methyl cellosolve as the internal standard. On the other hand, if both the analyte and the internal standard are fully evaporated (as is the case in the determination of the response factor for styrene versus methyl cellosolve: see Figure 6-2), the two plots are parallel, because now there is no longer any partitioning, and therefore the values of Q for both the analyte and the internal standard depend on the pressures, which are the same for both compounds.

The third case refers to *standard addition*, using MHE to estimate the total peak area values. Here the two plots will again be parallel, since analyte and matrix are the same in each sample. Plots A and C in Figure 6-5, representing the determination of residual toluene in a printed plastic film, illustrate such a case.

5.5.5 The Correlation Coefficient r

As discussed, the prerequisite of MHE measurements is the constancy of the area ratio Q in the whole concentration range. In practice, we do not calculate it for each step, but carry out linear regression analysis of the individual measurements and calculate the value of the area ratio Q from the slope of the regression plot q (cf. eqs. 5.25 and 5.26).

The problem with linear regression analysis is that it will always give a linear plot representing the best fit to the actual data points and the value of the slope of this "linear" plot. Therefore, it is important to know how good the fit is, and this is indicated by the correlation coefficient r. For an acceptable MHE measurement, the correlation coefficient should be at least 0.998. If it is less, one must investigate the spreading of the individual data points relative to the "linear" progression plot: whether there is a random variation, simply representing a weaker reproducibility of the analytical results (e.g., when there is only very little difference in the consecutive peak areas, approaching the precision of the measurement), or a definite trend in the plot one would obtain by actually connecting the individual data points. Such a situation is illustrated in cases F and G in Figure 5-15, indicating that we do not have a partition system. This is generally the case with solid samples representing adsorption systems. In Section 5.6 we shall deal with this question.

5.5.6 Evaluation of the Shape of the Regression Plots

Earlier we referred to the relative position (slope) of the linear regression plots (sample vs. calibration or analyte vs. standard), explaining the reasons for three characteristic situations: the usual case (slope of sample plot smaller than slope of standard), the case of the two plots being parallel, and the case of the slope of the sample plot approaching zero. In addition to these situations we can distinguish four more cases regarding the relative shape of the two regression plots: these represent incomplete equilibration, faulty headspace system, and various adsorption-type samples.

Figure 5-15 illustrates the seven shapes the plot representing the actual data points for the sample measurement can take and its relative position to the

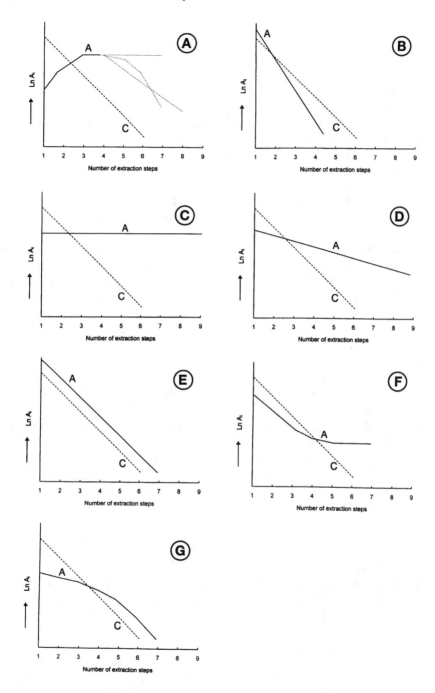

Figure 5-15. Various possibilities for the semilog relationship between peak area and the number of extraction steps in MHE measurements of a sample (*A*), compared with an external vapor standard (*C*, broken lines). For details, see text.

calibration plot. In all cases the use of an external vapor standard (obtained by TVT) is assumed. We briefly explain each case, indicating how the unsatisfactory situation may be corrected.

- **Case A:** The area values increase during the first MHE steps before they follow one of the possible shapes (dotted lines). This occurs because equilibration has not yet been established: the thermostatting time was too short. Therefore, the proper equilibration time must first be determined (e.g., by using the progressive working mode).
- **Case B:** The area ratio Q of the standard plot is higher than that of the sample plot; hence its slope is smaller. This should not occur, because as we have already derived, for a vapor standard Q should be smaller than in a partition system (cf. eq. 5.44 vs. eq. 5.43). This is a strong indication of a leaking system. If the septum is leaking after the first injection, the internal pressure (e.g., the water vapor pressure at vial temperature) causes a permanent flow from the pressurized vial to the atmosphere, carrying some analyte vapor with it. Therefore, the next MHE analysis gives a smaller peak than the external vapor standard which does not have such a problem.
- **Case C:** The plot for the analyte in the sample is practically parallel to the abscissa (i.e., the value of Q approaches unity). This situation is observed when the partition coefficient (more precisely: K/β) of the analyte is high. Such a pattern can usually be found with liquid samples, with a high solubility of the analyte. MHE is not the recommended method for quantitative analysis in such a case, and other methods (e.g., standard addition) should be preferred.
- **Case D:** This figure represents an "ideal" system for MHE; it is obtained when the standard and the analyte have different partition coefficients, or when an external vapor standard is compared with a partition system in which distribution is influenced by the solubility of the analyte in the matrix.
- **Case E:** In this case, the two plots are parallel. This means that the area ratio Q is the same for both the analyte and the standard. This is the situation for the case of total evaporation (TVT or FET) of both the analyte and the standard, when there is no longer any distribution between the two phases, or when the matrix effect is the same for both solutes (e.g., when standard addition is used).
- **Case F:** Connecting the area values, a nonlinear plot is obtained: in other words, there is no linear relationship between peak area and concentration. This is sometimes found in the case of FET where residual solid particles exercise strong adsorptivity that can be observed particularly at low analyte concentrations (see Figure 6-6). The same situation may occur with solid samples if the lower concentrations are adsorbed more strongly and released at a much slower speed. Usually the use of a displacer can correct the situation.
- **Case G:** Again, there is a nonlinear relationship between peak area and concentration. This is typical for solid samples representing adsorption

rather than partition systems (see Example 5.7). To solve the problem, the analyst must either use a displacer or transform the sample into a partition system. Another reason for such a deviation from linearity may be a nonlinear detector response for higher concentrations (see Section 9.7.1). In this case the sample should be diluted.

The conclusion of this discussion is that, particularly in the case of an unknown sample, an MHE investigation should always be the first step in establishing whether a linear system is present. Such an investigation can also provide information on the correctness of a quantitative method other than MHE. Just two examples:

1. We plan to apply standard addition to a solid sample, using gas phase addition, and we are interested to know whether the added analyte is properly distributed between the two phases. This can be established by carrying out MHE measurements with the original sample and with the sample to which additional analyte was added. If the added analyte is properly distributed between the two phases, the two regression plots will be parallel (case E), because the partition coefficient of the analyte is the same in either case. On the other hand, if the added analyte is not distributing properly between the two phases, the value of Q will be much smaller, approaching that of the external vapor standard.
2. We carry out external standard calibration using a simulated matrix. If the matrix in the two probes (sample and calibration) has the same effect, the partition coefficient of the analyte will be the same in both cases: thus, the two regression plots will again be parallel.

5.5.7 Influence of K/β

In a multiple headspace extraction measurement, our aim is to have as much difference between the peak areas of two consecutive measurements as possible. Since $A_{(i+1)} < A_i$ and $Q = A_{(i+1)}/A_i$, it is obvious that the value of Q should be as small as possible. We have already seen that in the case of two-phase partition systems (eq. 5.43), the value of Q depends on the values of K/β and ρ:

$$Q = \frac{K/\beta + \rho}{K/\beta + 1}$$

Considering ρ, we saw in eq. 5.40 that it is related to the pressures in the vial just before (p_h) and after (p_o) venting:

$$\rho = \frac{p_o}{p_h} \tag{5.40}$$

As mentioned earlier, the value of ρ lies around 0.3–0.6 and is fairly constant in most cases. Thus the magnitude of Q depends primarily on K/β, which can be varied more widely: if it is high, Q will approach unity.

Ioffe and Vitenberg in an earlier paper [8] recommended to keep the value of K/β close to unity: with $\rho = 0.5$, $K/\beta = 1.0$ would give $Q = 0.75$. This recommendation is, however, difficult to follow because there are practical limitations with respect to how far the phase ratio can be increased. We would suggest considering a practical limit of $K/\beta \approx 4$; assuming 1–10 mL as the practical range of sample volumes, this would represent a range of 21.3–1.23 for the phase ratio using a 22.3 mL vial, and thus a concomitant upper limit of $K = 85$ for the partition coefficient in MHE measurements [7,9].

These guidelines do not necessarily mean that one cannot carry out an MHE analysis with a sample if the value of K/β is above this proposed limit; however, it should be understood that accuracy will suffer. The partition coefficient of analytes with inherently high K values usually can be reduced by increasing the thermostatting temperature, and let us not forget that the K-versus-temperature relationship is exponential: a small increase in the temperature may result in a significant decrease of the partition coefficient (cf. Figure 2-4).

In the case of K values around 300 (e.g., at the determination of ethanol in aqueous solutions) the use of MHE is not recommended because the relative concentration of the analyte in the headspace is so small that the area differences between the two consecutive measurements would be just too little. In such a case the use of alternative calibration techniques, for example the internal standard or standard addition method is preferable. On the other hand, for solutions in which the analyte has a small partition coefficient (e.g., aqueous solutions of halogenated hydrocarbons [10]), the standard addition method suffers in accuracy, because the high volatility of the analytes may result in losses during sample handling. In such a case the MHE method is preferable.

Naturally, the phase ratio can be adjusted by the proper selection of the sample size. However, one should not forget that too small a sample volume represents problems in sample homogeneity, while a too large sample volume increases the time needed for equilibration. Figure 4-2 illustrated this for liquid samples.

Example 5.5″

It is intuitive to calculate the value of Q and K/β for the MHE determination detailed in Example 5.5. The headspace pressure p_h was about 234 kPa (abs) and $p_o = 100$; this gives a value of $100/234 = 0.427$ for ρ. The value of q was established from the linear regression analysis as -0.83304 (cf. Table 5-5); this gives $Q = 0.4347$. Thus,

$$\frac{K/\beta + 0.427}{K/\beta + 1} = 0.4347$$

From this we can calculate that $K/\beta = 0.014$, which is well below the recommended limit. With the sample volume of 0.70 mL in the 22.3 mL vial ($\beta = 30.86$), we get a partition coefficient of $K = 0.42$ for EO in PVC and such a low value corresponds practically to FET.

5.6 Analysis of Solid Samples (Adsorption Systems)

We have discussed in detail the questions associated with solid samples in Section 4.1.3. A direct analysis is possible if the sample can be treated as a *partition system*: this is the case when, for example, a polymer sample is heated above its glass transition temperature. In such systems the distribution (partition) coefficient may be considered (assuming dilute solutions) to be constant and independent of analyte concentration. The MHE method can serve as a quick test of whether this is the case: if the plot of $\ln A_i$ versus $(i - 1)$ is linear, a partition system can be assumed. On the other hand, nonlinearity of the MHE plot indicates that the distribution coefficient is not independent of concentration and/or that adsorption effects are present. MHE plots reflecting such a situation were illustrated in Figure 5-15 (cases F and G).

If an *adsorption system* is present, it must be converted into a system in which the surface properties are altered. In many cases even a small amount of an added liquid can act as a *modifier* (*displacer*), changing the surface into a weak adsorption system with homogeneous adsorptivity, extending the linear concentration range, and releasing the adsorbed analyte molecules. When a thin layer of the modifier is formed on the surface, the solid sample may still appear dry: however the surface was already changed into a partition system. We call this mode of operation the *surface modification method*.

When the amount of liquid displacer added to the sample is increased further, it begins to separate from the solid sample and build up a separate liquid phase into which the displaced analyte molecules are eluted: they will now partition between the headspace of the vial and this new liquid phase. The solid sample particles remain suspended in this liquid, but usually they have no further influence, and the added solvent is now the matrix (*suspension approach*).

Both the surface modification and suspension approaches require the volatile analytes to be superficially adsorbed, and thus accessible to the liquid displacer. If, however, their molecules are occluded in the crystalline structure and cannot be released by diffusion, both approaches will fail. In this case the matrix must be destroyed either by melting or by dissolution. To avoid such uncertainty, some official methods (e.g., the U.S. Pharmacopeia [11]) suggest in general the "solution approach" (see Section 4.2).

5.6.1 Suspension Approach

In an adsorption system the distribution of the analyte between the surface of the adsorbent and the gas phase in the headspace vial often depends on the concentration, and this is obvious if the relationship of $\ln A_i$ to $(i - 1)$ in the MHE analysis is found to be nonlinear. The underlying adsorption effects may be overcome by using a displacer that has a stronger affinity to the solid (adsorbent) than the analyte, thus accelerating the desorption of the analyte.

The desorption and thus the extraction yield are further supported by an increasing volume of the liquid displacer similar to any other extraction procedure. The solid sample, therefore, may be suspended in a solvent that will extract the analyte from the surface of the solid sample. The resulting suspension is thus identical to a simple liquid solution and can be treated as such.

The solid particles of the suspended sample have no further influence on the gas/liquid partitioning process of the dissolved analyte in the headspace vial, and the calibration technique is straightforward: the liquid displacer is the matrix, and this solvent is used to prepare an external standard (cf. Sections 4.5 and 5.3). Any other calibration technique, internal standard, or standard addition can also be applied. MHE, however, is of restricted use here, because in general the solvent selected has too good a solubility for the displaced analyte.

The completeness of desorption by the *suspension approach* is illustrated in the use of adsorbent tubes to monitor the volatile pollutant content of air. In these tubes, which contain a strong adsorbent such as charcoal, the collection of volatile compounds from air is carried out either by pumping a certain volume through them (*active sampling*) or, without the use of a pump, by diffusive sampling (*passive sampling*) (see, e.g., ref. 12). The official methods dealing with determination of the adsorbed amounts of pollutants use extraction with an organic solvent such as carbon disulfide to recover the adsorbed compounds. However, there is a more pleasant and less toxic alternative, based on HS-GC, since complete desorption may also be accomplished in a headspace vial, by adding a displacer (e.g., benzyl alcohol, benzoic acid esters, or dimethylacetamide) having a stronger affinity to charcoal than the compounds of interest. These displacers have the following particular advantage: on almost all GC columns, their retention time is longer than that of the pollutant to be determined, and thus the displacer's peak will not interfere with the peaks of the analytes. Also, analysis time can be speeded up by backflushing the column after the pollutants have been eluted.

The use of such a displacer is illustrated here by the following experiment. A small volume (1 µL) of a mixture of four compounds was added into two standard headspace vials, where it totally evaporated. The first vial was empty, while the second contained 400 mg of activated charcoal. The headspace of both vials was analyzed in the usual way, and Figure 5-16 shows the two resulting chromatograms. As seen, charcoal totally adsorbed the organic compounds: the chromatogram only shows a straight baseline. The third vial repeated vial 2: the 1 µL mixture was added to 400 mg charcoal. However, after complete adsorption had been achieved, 2 mL of benzoic acid benzyl ester was added into the vial and, after a 1-hour equilibration at 80 °C, the headspace of the vial was analyzed as in the other two cases. Vial 4, which already contained 2 mL benzoic acid benzyl ester and was fortified with 1 µL of the solution of the four components, was used as an external calibration standard. By comparing the chromatogram with that obtained from vial 3, the yield of

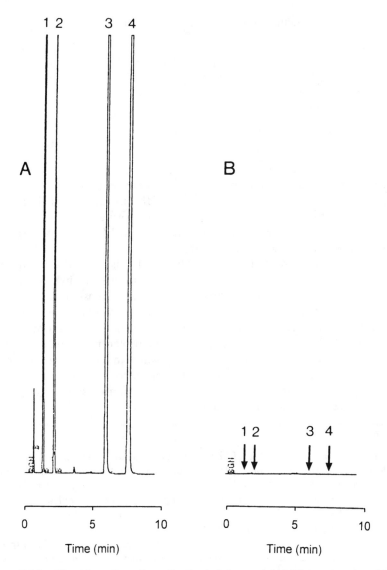

Figure 5-16. The adsorption effect of activated charcoal. (A) Chromatogram obtained when vaporizing 1 μL of a mixture in a headspace vial, at 80 °C. (B) The same sample as in (A), but adding 400 mg activated charcoal into the headspace vial; positions at which the four components would elute are marked by arrows. *GC conditions*: 25 m × 0.32 mm I.D. fused-silica, open-tubular column coated with bonded phenyl (5%) methyl silicone stationary phase; film thickness: 0.3 μm. Column temperature: 80 °C. Flame-ionization detector with attenuation of ×8. *Peaks*: *1* = methyl ethyl ketone, *2* = 1,1,1-trichloroethane, *3* = toluene, *4* = *n*-octane.

desorption can be calculated: the results are listed as follows:

Compound	recovery (%)
1,1,1-trichloroethane	102
methyl ethyl ketone	93
toluene	104
n-octane	110

These data show that the desorption was essentially complete.

Figure 5-17 shows a practical application of this method: the determination of tetrachloroethylene from a personal air monitoring tube. Here 1 mL of benzyl alcohol was used as the displacer: the obtained C_2Cl_4 peak corresponded to an air concentration of 235 mg/m³.

The surface of charcoal is hydrophobic, and this was why we used an organic solvent as the displacer. Most natural samples are hydrophilic, and with them, water should be the first choice. The suspension approach was originally developed for the determination of water in solid samples that are not soluble in the solvents used [13], but for this analysis it is obvious that another polar solvent must replace water as the displacer. Methyl cellosolve is a good solvent for this purpose, but other water-miscible solvents such as dimethylacetamide or dimethylformamide may also be used. Example 5.6, illustrating the determination of the water content of an instant soup powder, utilized an internal standard for calibration.

Example 5.6

Sample amount was 500 mg of the solid material to which 2.0 mL of methyl cellosolve (2-methoxyethanol, Riedel-de Haën, water content max 0.1%) was added containing 1% of methanol used as the internal standard; its amount present was 15.83 mg. The vial was equilibrated at 110 °C for 60 minutes. GC analysis was carried out with a thermal-conductivity detector. The calibration factor for methanol/water was determined separately as $f_i = 5.854$.

Figure 5-18 shows the chromatogram obtained for the sample. The respective peak areas were 195 counts (water) and 695 counts (methanol). The amount of water present is (cf. eq. 5.8):

$$15.83 \times 5.854 \times 195/695 = 26.00 \text{ mg}$$

Thus the water concentration is 5.20 wt%.

However, this result must further be corrected for the water blank, which is caused by the inherent humidity of the air filling the original headspace of the vial. This is an important point, which must be considered when analyzing the water content of a sample. There are two possibilities. At first the amount of water in the headspace vial can be taken from the water vapor density (μg/mL) in Table 2-1 as a function of the ambient temperature and the relative humidity. On the day of this analysis the temperature in the laboratory was about 20 °C (water density:

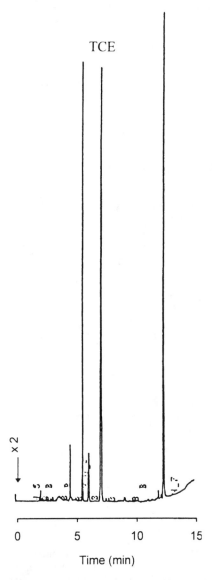

Figure 5-17. Headspace analysis of the activated charcoal filling of a personal monitoring tube for tetrachloroethylene (*TCE*). Displacer: benzyl alcohol. *HS conditions*: Sample: charcoal from a personal monitoring tube by passive sampling for 8 hours to which 1 m of benzyl alcohol was added. Equilibration at 120 °C for 30 minutes. *GC conditions*: Two 25 m × 0.32 mm I.D. fused-silica, open-tubular columns coated with bonded phenyl (5%) methyl silicone stationary phase; film thickness: 0.3 μm. Backflush configuration. Column temperature: 80 °C for 10 minutes, then programmed at 15 °C/min to 150 °C. Flame-ionization detector.

Figure 5-18. Headspace analysis of water in an instant soup powder, using the suspension approach (Example 5.6). *HS conditions*: Sample: 500 mg dry soup powder + 2.0 mL methyl cellosolve (2-methoxyethanol), containing 1% methanol as internal standard. Equilibration at 110 °C for 60 minutes. *GC conditions*: 50 m × 0.32 mm I.D. fused-silica, open-tubular column coated with bonded cyanopropyl (14%) methyl silicone stationary phase; film thickness: 1 μm. Column temperature: 70 °C. Thermal-conductivity detector. Carrier gas: Helium, 3.5 mL/min; makeup gas (He): 19.0 mL/min. *Peaks*: *1* = air; *2* = water, 5.16 wt%; *3* = methanol.

17.3 μg/mL) and the relative humidity 60% (measured with a notoriously unprecise hair hygrometer). The approximate volume of the sample + solvent was 2.5 mL. Therefore the total amount of water in the remaining 19.8 mL gas volume in the 22.3 mL vial was 17.3 × 0.6 × 19.8 = 206 μg or 0.206 mg.

The second possibility is to determine the water blank by a separate measurement, say by standard addition of water into an empty vial [13]. A blank of 0.215 mg was found at 20 °C for the empty 22.3 mL vial that corresponds to a relative humidity of 56%:

$$\frac{215}{22.3 \times 17.3} = 0.56$$

The corresponding water blank in the 19.8 mL headspace volume is 17.3 × 0.56 × 19.8 = 192 μg or 0.192 mg.

Applying these corrections, the amount of water in the instant soup powder sample would be

$$26.00 - 0.206 = 25.794 \text{ mg and } 26.00 - 0.192 = 25.808 \text{ mg}$$

and its concentration in the sample can be established in both cases as 5.16%, representing a difference of only 0.8%.

5.6.2 Surface Modification Techniques

Whenever a solid sample fulfills the presupposition of the suspension approach, it is in fact the simplest sample type in HS-GC, since no matrix effect other than that of the pure solvent has to be taken into account. However, the extraction into a liquid solvent automatically dilutes the desorbed analyte and reduces headspace sensitivity. Another problem with organic solvent is the content of impurities which often interfere with the analytes in the chromatogram. Both effects can be minimized if the amount of the liquid displacer is continuously reduced. To achieve the desired displacement effect, it should in fact suffice to reduce the amount of the liquid displacer until only the surface of the solid sample is covered. Such a small amount of a liquid displacer is now better described by the term "modifier" rather than by "displacer" [14].

As an example for this effect, we show the determination of 1.9% isopropyl alcohol and 5.2% water in a drug powder [6]. First we tried to analyze a dry sample: however, MHE plot A in Figure 5-19 corresponded to the case F type (Figure 5-15), indicating a nonlinear behavior in the low concentration region due to increasing adsorptivity for low concentrations. As a next step we used the suspension approach, adding 2 mL ethyl cellosolve to 100 mg of the drug powder. However, as can be seen in the chromatogram (Figure 5-20A), only a very small peak was obtained for isopropyl alcohol (peak 2). Therefore, we tried the surface modification approach, adding 307 mg glycerol to 100 mg of the sample.* Due to the high porosity of the powder, the amount of glycerol

---

* Because we also wanted to determine the water content of the sample in the same analysis, we had to use a thermal conductivity detector, with a makeup gas. For this reason we obviously could not use water as the displacer. Therefore, glycerol was selected finally as a less volatile solvent.

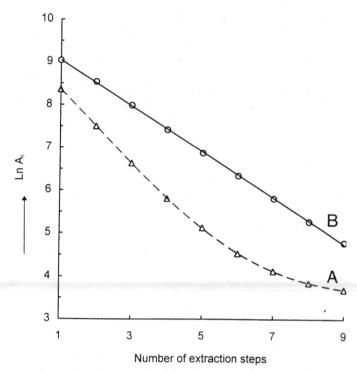

Figure 5-19. Regression plots for the MHE determination of isopropyl alcohol in a drug powder. A = 100 mg dry powder, thermostatted 90 minutes at 110 °C, B = 100 mg drug powder + 307 mg glycerol, thermostatted 60 minutes at 110 °C (correlation coefficient r = 0.99995).

was just sufficient to wet the surface: the sample still looked dry. As shown in Figure 5-20B, a much better sensitivity was obtained in this way, and the MHE plot showed a straight line (plot B, Figure 5-19).

The detrimental effect of surface adsorption and changes in the humidity of the sample can be illustrated on the example of determining residual *trans*-1,2-dichloroethylene in instant coffee [15]. This solvent had been used in the industrial decaffeination process. Upon analyzing the dry sample with the MHE method, a nonlinear plot of ln A_i versus $(i + 1)$ was obtained (see Figure 5-21A) because of the combined effect of adsorption and changes in the concentration of water in the sample.* Thus one can conclude that the value of Q is not constant. The problem can be solved by adding an excess of water to the sample: Figure 5-21B — which is linear — was obtained in this way. The pertinent data are given in Example 5.7.

---

* Coffee powder always has some humidity, but the water peak cannot be seen when a flame-ionization detector is used in the GC. However, with a thermal-conductivity detector, the MHE method can actually be used to determine the water content of instant or roasted coffee [16].

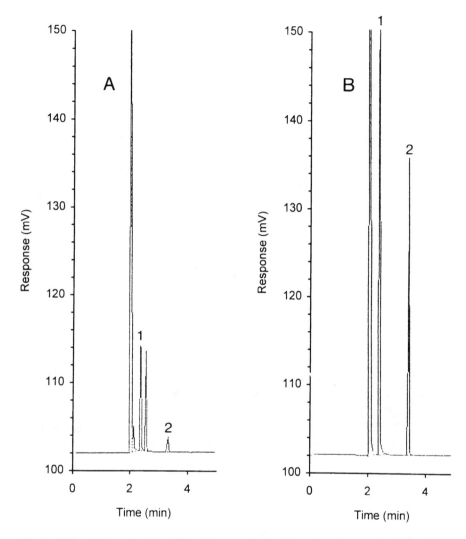

Figure 5-20. Determination of the isopropyl alcohol and water content of a drug powder, using (A) the suspension approach and (B) the surface modification method. *HS conditions*: Sample: 100 mg powder. Displacer: (A) ethyl cellosolve (2.0 mL), (B) glycerol (300 mg). Equilibration at 110 °C for 60 minutes. *GC conditions*: 50 m × 0.32 mm I.D. fused-silica, open-tubular column coated with bonded cyanopropyl (14%) methyl silicone stationary phase; film thickness: 1 μm. Column temperature: 60 °C. Thermal-conductivity detector. Carrier-gas: Helium, 3.5 mL/min; makeup gas (He): 19.0 mL/min. *Peaks: 1* = water, 5.2%; *2* = isopropyl alcohol, 1.9%.

Source: Reproduced by permission from *Pharmacopeial Forum* [6]. Copyright 1994, The USP Convention, Inc.

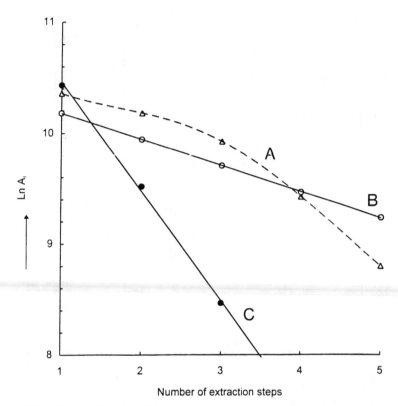

Figure 5-21. Regression plots for the determination of *trans*-1,2-dichloroethylene in instant coffee by MHE [17]. *A* = 470 mg dry coffee powder, *B* = 470 mg dry coffee powder + 100 μL water, *C* = external calibration standard (3 μL of a 5.06 mg/mL solution in dioxane). Equilibration at 80 °C for 30 minutes. For details see Example 5.7.

Example 5.7

Into a headspace vial 470 mg coffee powder and 100 μL water were added. The calibration standard consisted of a 5.06 mg/mL solution of *trans*-1,2-dichloroethylene in dioxane; 3 μL of this solution (containing 15.18 μg dichloroethylene) was put into an empty vial, which was carried through the same measurements. In a third vial 470 mg of coffee powder with no added water was also analyzed. All three vials were equilibrated at 80 °C for 30 minutes.

A 50 m × 0.25 mm I.D. fused-silica, open-tubular column coated with SE-54 phenyl (5%) vinyl (1%) methyl silicone stationary phase was used in the gas chromatograph, at 70 °C. Table 5-6 lists the results of the MHE measurements, together with the regression data; the corresponding plots are shown in Figure 5-21.

The following Σ *A* values were obtained:

coffee powder + water:	125,096
calibration standard:	55,442

Table 5-6 Determination of residual *trans*-1,2-dichloroethylene in instant coffee powder; using MHE measurement (Example 5.7)*

	Peak area (counts) for:		
i	Dry coffee powder	Coffee powder + water	Calibration standard
1	31,510	26,376	33,931
2	26,343	20,777	13,585
3	20,398	16,385	4,742
4	12,396	12,899	
5	6,641	10,274	
Linear regression			
Correlation coefficient r	−0.970309	−0.99996	−0.999190
Slope q	−0.38679	−0.23624	−0.98393
$Q = e^{-q}$	0.6792	0.7896	0.3738
Intercept A_1^*	36,712	26,321	34,716
Total area ΣA		125,096	55,442

* For the corresponding regression plot, see Figure 5-21.

The total amount of dichloroethylene in the sample was

$$\frac{125,096}{55,442} \cdot 15.18 = 34.25 \, \mu g$$

corresponding to a concentration of 0.00729 wt% or 72.9 ppm.

In this calculation we did not correct for sample volume (see Section 5.5.4). If we assume that the volume of the coffee powder in vial 1 is 0.5 mL, then the volume correction is $f_V = 22.3/21.8 = 1.023$ and the value of ΣA_{st} would be 56,714; in this way the total amount of dichloroethylene in the sample would be 33.48 μg, corresponding to a concentration of 71.2 ppm.

Sometimes the addition of a very small amount of water can result in dramatic changes, mainly in the time needed for equilibration. A good example for this is the determination of small amounts of ethanol and dichloromethane in a pharmaceutical drug powder, the matrix of which consisted of carbohydrates [17]. Figure 5-22A shows the equilibration of the original (dry) sample, while Figure 5-22B represents the addition of 5 μL of water to the 200 mg of powder. For these plots a series of samples were analyzed by HS-GC using the progressive working mode, and the resulting peak heights were plotted against the thermostatting time. In the case of the original (dry) sample, dichloromethane equilibrated very quickly, in less than 20 minutes; however, with the polar ethanol, no equilibrium could be reached even after 100 minutes. This slow release of ethanol from the sample matrix can be explained by strong interaction with the hydroxyl groups of the carbohydrates representing the matrix. The addition of only 5 μL of water dramatically changed the situation: as a more polar substance, it displaced the ethanol molecules on the surface of

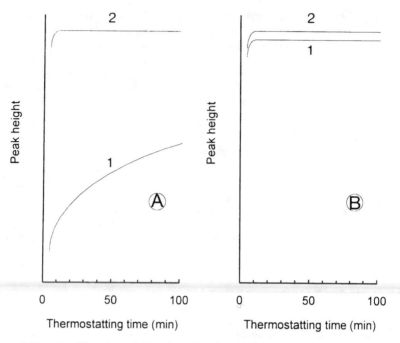

Figure 5-22. Equilibration of (1) ethanol and (2) dichloromethane from a dry pharmaceutical powder. (A) Original dry sample (200 mg). (B) Sample (200 mg) + 5 μL water.

the solid sample. Thus the ethanol equilibrated as rapidly as dichloromethane, and in the state of equilibrium the ethanol peak was much higher, while in the case of the dry sample it was only a fraction of the correct peak height.

For the actual MHE measurements, 5 μL of water was added to 200 mg of the original sample. An external vapor standard was prepared using the TVT, adding 0.8 μL of a solution of ethanol and dichloroethane in toluene (with a concentration of 20 μg/μL of each compound) into an empty headspace vial. Equilibration was carried out at 60 °C for 20 minutes. The regression plots show excellent linearity ($r = 0.9999+$). The measured concentrations were 84.7 ppm for ethanol and 257 ppm for dichloromethane. The measurements are outlined in Example 5.8 for ethanol only.

Example 5.8

The amount of the pharmaceutical drug analyzed was 200 mg to which 5 μL water was added; equilibration took 20 minutes at 60 °C. A 50 m × 0.32 mm I.D. fused-silica, open-tubular column coated with Carbowax 1000 poly(ethylene glycol) was used at 70 °C in the GC. An 18.8 mg/mL solution of ethanol in toluene was used as the calibration standard; 0.8 μL of this solution (containing 15.04 μg of ethanol) was added to an empty sample vial, where it totally evaporated. Table 5-7 lists the results

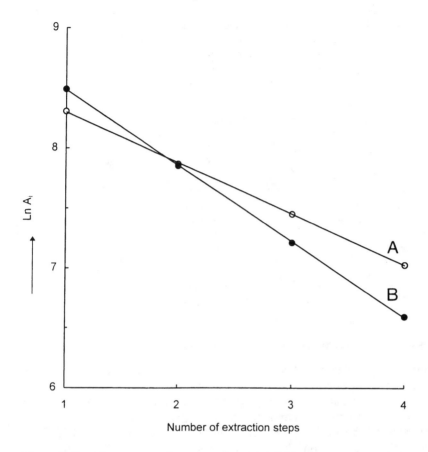

Figure 5-23. Linear regression plots for the MHE measurements aiming at the determination of residual ethanol in a pharmaceutical drug powder (Example 5.8). A = wetted drug powder, B = external vapor standard.

of the MHE measurement, together with the regression data; the corresponding plots are shown in Figure 5-23.

The calculated values of the sum of the peak areas are:

powder + 5 μL water: 11,636

calibration standard: 10,346

The amount of ethanol present is

$$\frac{11,636}{10,346} \cdot 15.04 = 16.915 \, \mu g$$

which corresponds to a concentration of 0.00846 wt% or 84.6 ppm.

A two-point calculation gives a total amount of 16.246 μg and a concentration of 0.00812 wt% (81.2 ppm).

Table 5-7 Determination of traces of ethanol in a pharmaceutical drug powder, using MHE measurement (Example 5-8)

	Peak area (counts) for:	
i	Powder $+5\,\mu L$ water	Calibration standard
1	4055	4876
2	2608	2566
3	1721	1357
4	1130	732
Linear regression		
Correlation coefficient r	−0.999915	−0.99956
Slope q	−0.42489	−0.63260
$Q = e^{-q}$	0.6538	0.5312
Intercept A_1^{*}	4,028	4.850
Total area ΣA	11,636	10,346

* For the corresponding regression plots, see Figure 5-23.

5.6.3 Highly Adsorptive Solid Samples

An important and generally unknown behavior of highly adsorptive solid samples entails the speed and extent of desorption. Preliminary investigations are needed to determine the recovery with the particular desorption technique. For this purpose artificial standards must be prepared by spiking the pure adsorbent with the analytes or surrogates. It has been recognized that the recovery also depends on the time that elapsed between adsorption and desorption. Very often, such artificial standards are prepared, and soon afterward recoveries near 100% are determined, while, on the other hand, lower recoveries are found if the time between spiking the samples and its analysis by desorption is extended.

For example, an artificial soil standard was prepared according to the proposal of Hellmann [18], from 1 g of Fuller's earth, a strong adsorbent, previously dried at 200 °C and, subsequently equilibrated with the humidity of air at room temperature. When the standard was spiked with volatile aromatic and chlorinated hydrocarbons (each compound at 2.5 µg/g), and then desorbed by the addition of 2 mL of water as the displacer, at 60 °C, the speed of equilibration was found to depend on the elapsed time between spiking and addition of the desorber. If desorption was carried out 3 hours after spiking the solid with the hydrocarbons, their recovery was 91% after thermostatting for 1 hour at 60 °C. On the other hand, if desorption was carried out 2 days after spiking the sample (which had been stored at room temperature), then after 1 hour of thermostatting at 60 °C, the recovery was only 21%, and 4 hours of thermostatting was needed to achieve 91% recovery [19,20]. For comparison, we note that an aqueous solution of the same hydrocarbons will fully equilibrate at 60 °C within 45 minutes. This difference in desorption speed is attributed to a slow migration of the adsorbed molecules into the micropores of the solid after adsorption, and a slow reverse process during desorption.

As a solution to this problem with particular respect to the case of highly adsorptive soil samples (not all are adsorptive; e.g., sand and wet clay are not adsorptive at all), it was proposed to carry out a two-stage process in the analysis of soil samples [19,20]. The displacement process should be carried out with water, soon after sample collection, by storing the capped vials containing the water slurry of the soil sample (typically, 1 g of soil with 2 mL of water) in an oven, at 95–100 °C, for a few hours (e.g., overnight). When finished, the desorption is not reversible: from then on, the volatile organic compounds remain in the aqueous phase and the soil is now present just as an inactive sediment. Therefore, from now on, the sample vials can be stored at lower temperature and analyzed later, at a lower temperature, with the usual equilibration time (e.g., 45 minutes at 60 °C).

This procedure is contrary to the usual recommendations that the soil samples be stored in a refrigerator between collection and actual analysis. However, our experience [19,20] has shown its superiority to the established methods.

The importance of this observation is that whenever ad- and desorption processes are involved — highly adsorptive samples need longer desorption times than generally expected — the equilibration between the liquid displacer and the headspace is always much faster. Such an example was given in Example 5.8 (Figure 5-22). The use of a shaker does not help much, since mechanical shaking apparently has no effect on the micropore diffusion. Ultrasonification is more effective here and can be applied if in the recommended two-step procedure the desorption is carried out in a liquid bath with ultrasonifaction at high temperature (near 100 °C).

This procedure, where the slurry of the soil sample is desorbed at a temperature near 100 °C immediately after collection, also helps to overcome problems with microbial degradation of volatile aromatic and halogenated hydrocarbons. In this way the sample is automatically pasteurized at the high temperature, making the addition of chemicals unnecessary.

Soil samples with a significant amount of organic material are highly adsorptive (cf. later discussion for Figure 9-8) and usually lower recoveries (60–80%) are found [19] even with this two-step procedure. Apparently in addition to physical ad- and desorption processes, chemical reactions (with humic acids?) cannot be excluded at all.

5.7 Calibration Techniques with Headspace Samples of Varying Volumes

As discussed earlier, it is always desirable — and sometimes (in the case of analytes with low partition coefficients) absolutely necessary — to measure the same volumes of samples and standards into the vials. However, in the case of viscous samples containing highly volatile analytes, where quick sample handling is crucial, it is difficult to exactly reproduce the volume. One solution is to establish the concentration of the analyte in the sample by indirect calculation knowing the individual sample volumes. This is based on eq. 2.19, the fundamental equation of HS-GC:

$$A \propto \frac{C_o}{K + \beta} \qquad (2.19)$$

which we now write in the following way:

$$C_o = f \cdot A \cdot (K_i + \beta) \qquad (5.48)$$

where K_i is the partition coefficient of the analyte in the sample matrix, β is the phase ratio of the sample in the vial, A is the peak area obtained, C_o is the original concentration of the analyte in the sample, and f is a proportionality factor. It should be mentioned that this calibration technique requires an identical value of the partition coefficient K in both the sample and the calibration standard. This can be achieved only if the pure matrix is available for preparing an external standard.

We can write this relationship for the sample (s) and the external calibration standard (st) in the following way:

$$C_{o,s} = f \cdot A_s \cdot (K_i + \beta_s) \qquad (5.49a)$$
$$C_{o,ex} = f \cdot A_{ex} \cdot (K_i + \beta_{ex}) \qquad (5.49b)$$

In eqs. 5.49 we have two equations with three unknowns: $C_{o,s}$, K_i, and f_i. To solve them for $C_{o,s}$, one of the other two unknowns must be established. The easiest way is to determine the value of the partition coefficient, using any of the ways to be discussed in Chapter 9. Knowing this value, $C_{o,s}$ can be determined:

$$C_{o,s} = C_{o,ex} \cdot \frac{A_s}{A_{ex}} \cdot \frac{K + \beta_s}{K + \beta_{ex}} \qquad (5.50)$$

As an illustration of this calculation, we utilize the data given later (Example 9.9), the determination of the partition coefficient of methyl ethyl ketone (MEK) using the PRV MHE method (see Section 9.4.3.2), recalculating the concentration of MEK in the 5-mL sample.

Example 5.9

The volumes of the sample and the standard were 5.0 and 1.0 mL, respectively. The vial volume was 22.3 mL; thus the phase ratio values are $\beta_s = 3.46$ and $\beta_{ex} = 21.3$. The concentration of MEK in the standard was 2.415 mg/mL, and the partition coefficient of MEK in the air–water system at 70 °C was $K = 44.5$; for other details, see Example 9.9. The following peak area values were obtained: $A_s = 1,470$; $A_{ex} = 5,346$. Thus:

$$C_{o,s} = 2.415 \cdot \frac{1470}{5346} \cdot \frac{44.5 + 3.46}{44.5 + 21.3} = 0.484 \text{ mg/mL}$$

As indicated in Example 9.9, the amount of MEK in the 5.0 mL solution was 2.415 mg (3 μL), which gives a concentration of 0.483 mg/mL. Thus, the agreement is excellent.

Markelov [21] has developed a special method for quantitative headspace analysis called the *variable volume technique*, in which the volume of the sample, and thus the phase ratio β is changed in several vials, with the effect that the underlying partition coefficient K is automatically determined and included in the calibration procedure in the same way as with the PRV technique (see later: Section 9.4.2).

5.8 Analysis of Gas Samples

As discussed in Section 4.3.1, the headspace vial can also be used to collect gas samples, and we presented two simple methods of execution. The first utilized a small hand pump, while in the second case, the gas to be analyzed was blown into the vial. We now give examples for each technique.

In the first case, the atmosphere of an underground parking garage was analyzed for trace aromatic hydrocarbons (BTEX: benzene, toluene, ethylbenzene, and xylenes) [22]. The vial was filled with atmospheric air using a small hand pump as shown in Figure 4-6, and cryofocusing (see Section 3.8) was used to further improve the detection limit. With 3-minute sample transfer to the cryogenic trap and using a flame-ionization detector, the detection limit is about 10 ppb; however it could be further lowered by increasing the transfer time and using a photo-ionization detector. The external vapor standard consisted of a BTEX solution in propylene glycol carbonate, fully evaporated in the vial.

Example 5.10

Figure 5-24 shows a typical chromatogram; the instrumental and analytical conditions are given in the caption.

*Calibration by external vapor standard.** Stock solution II was used, prepared as described in Example 4.1; its benzene concentration was 8.79 μg/mL. A 5-μL aliquot of this solution (containing 43.95 ng of benzene) was added to an empty headspace vial of 22.3 mL volume, where it evaporated: the total amount of benzene in the vial was 43.93 ng, and the benzene concentration in the vial's gas volume was 1.97 ng/mL.

The following peak area values were obtained:

sample: 18,343 counts

calibration standard: 44,681 counts

Thus we have in the actual gas sample (22.3 mL):

$$\frac{18,343 \times 43.93}{44,681} = 18.035 \text{ ng benzene}$$

and its concentration is 809 μg/m^3.

The second example concerns the analysis of the carbon monoxide content in the breath of a smoker and a nonsmoker [23]. Here the technique discussed in Section 4.3.1 was applied for sample collection, blowing the breath

---

* Calculation is given for benzene only. For the measured concentrations of the other aromatic hydrocarbons, see the caption of Figure 5-24.

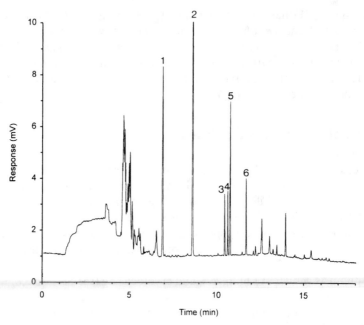

Figure 5-24. Analysis by cryofocusing of the atmosphere of an underground parking garage for BTEX (Example 5.10). *HS conditions*: Vial sampling technique; vial equilibrated at 80 °C for 35 minutes; transfer for 3 minutes into a cryogenic trap (55 cm × 0.32 mm I.D. fused-silica, open-tubular column coated with methyl silicone phase; film thickness: 1 μm), with a temperature gradient between −80 °C and −20 °C. *GC conditions*: 60 m × 0.25 mm I.D. fused-silica, open-tubular column coated with bonded poly(propylene glycol) stationary phase; film thickness: 0.25 μm. Column temperature: 1 minute isothermal at 40 °C, then programmed at 20 °C/min to 65 °C; 4 minute isothermal at 65 °C, then programmed at 10 °C/min to 120 °C. Flame-ionization detector. *Peaks* (concentration): *1* = benzene (809 μg/m³), *2* = toluene (1596 μg/m³), *3* = ethylbenzene (228 μg/m³), *4* = *p*-xylene (228 μg/m³), *5* = *m*-xylene (531 μg/m³), *6* = *o*-xylene (245 μg/m³).

Source: Reproduced with permission of *LC/GC International* [22].

sample into the vial through a plastic tube. To increase sensitivity, the carbon monoxide content of the breath* was transformed during analysis into methane by hydrogenation

$$CO + 3H_2 \rightarrow CH_4 + H_2O$$

---

* Naturally, the carbon dioxide content of breath is also transformed into methane:

$$CO_2 + 4H_2 \rightarrow CH_4 + 2H_2O$$

Since CO and CO_2 were already separated in the column prior to entering the reaction tube, they will appear as separate peaks in the chromatogram. However, when the column is used in the backflush mode, the large peak corresponding to carbon dioxide can be eliminated before it would have reached the reactor.

on a nickel catalyst, permitting the use of a flame-ionization detector. This reaction was first described in 1902 by Sabatier and adapted to gas chromatography in 1961–1962 [24,25]. In practice, a small reactor tube containing the catalyst is inserted by the column outlet and the jet of the FID and the hydrogen flow needed for the detector is introduced just upstream of the reactor tube. Today such "methanizer" systems represent routine accessories of gas chromatographic instruments (see, e.g., ref. 26). For calibration, an external standard sample was prepared by injecting 5 μL of pure carbon monoxide gas with a gas-tight syringe into an empty headspace vial.

The two chromatograms obtained are shown in Figure 5-25. The "methane" peak corresponds to the original methane content of breath. The carbon

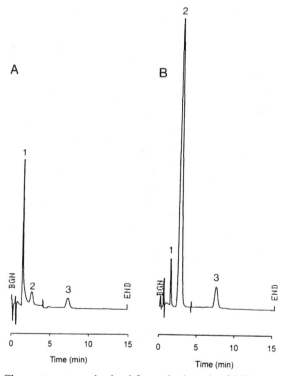

Figure 5-25. Chromatograms obtained from the breath of (A) a nonsmoker and (B) a smoker, by headspace–gas chromatography, transforming the CO (and CO_2) content of the breath into methane, using a methanizer and a flame-ionization detector [23]. *HS conditions*: Vial temperature: 40 °C. *GC conditions*: Two 50 cm × 1/8 in. O.D. packed columns containing Carbosieve SII, 60/80 mesh; backflush mode. Column temperature: 40 °C. Carrier gas: Helium, 33 mL/min. Flame-ionization detector with attenuation of × 8. Hydrogen inlet pressure to the reactor: 131 kPa. *Peaks: 1* = "air" (disturbance caused by the humidity in the breath plus formation of water by the reaction of oxygen + hydrogen over the catalyst), *2* = carbon monoxide (converted to methane), 3 ppm in chromatogram A (nonsmoker), 70 ppm in chromatogram B (smoker), *3* = methane (originally present in breath).

monoxide concentrations estimated by comparison with the external standard were 3 ppm for the nonsmoker and 70 ppm for the smoker.

References

1. G. Machata, *Mikrochim. Acta*, **1964**, 262–271.
2. G. Machata, *Blutalkohol*, **4**, 3–11 (1967).
3. G. Machata, *Clin. Chem. Newsl.* **4**, 29–32 (1972).
4. D. J. Brown and W. C. Long, *J. Anal. Toxicol.* **12**, 279–283 (1988).
5. B. Kolb, *J. Chromatogr.* **122**, 553–568 (1976).
6. B. Kolb, *Pharmacopeial Forum*, **20**, 6956–6960 (1994).
7. B. Kolb and L. S. Ettre, *Chromatographia*, **32**, 505–513 (1991).
8. B. V. Ioffe and A. G. Vitenberg, *Chromatographia*, **11**, 282–286 (1978).
9. L. S. Ettre and B. Kolb, *Chromatographia*, **32**, 5–12 (1991).
10. B. Kolb, M. Auer, and P. Pospisil, *J. Chromatogr.* **279**, 341–348 (1983).
11. *U.S. Pharmacopeia XXIII.* Organic Volatile Impurities (467), Method IV: 1995; pp. 1746–1747.
12. F. Bruner, *Gas Chromatographic Environmental Analysis*, VCH Publishers, New York, 1993; pp. 119–179.
13. B. Kolb and M. Auer, *Fresenius Anal Chem* **336**, 291–296, 297–302 (1990).
14. N. Onda, A. Shinohara, H. Ishi, and A. Sato, *HRC*, 357–360 (1991).
15. B. Kolb, P. Pospisil, and M. Auer, *Chromatographia*, **19**, 113–122 (1984).
16. B. Kolb, in J. Gilbert (editor), *Analysis of Food Contaminants*, Elsevier, Amsterdam, 1984; pp. 117–156.
17. B. Kolb, in R. A. A. Maes (editor), *Topics in Forensic and Analytical Toxicology*, Elsevier Science Publishers B.V., Amsterdam, 1984; pp. 119–126.
18. H. Hellmann, *Fresenius Z. Anal. Chem.* **327**, 524–529 (1987).
19. T. C. Voice and B. Kolb, *Environ. Sci. Technol.* **27**, 709–713 (1993).
20. B. Kolb, C. Bichler, M. Auer, and T. C. Voice, *HRC*, **17**, 299–302 (1994).
21. M. Markelov, D. Mendel, and L. Talanber, in Pittsburgh Conference Abstracts, 1983, No. 206.
22. B. Kolb, *LC/GC Int.* **8**, 512–524 (1995).
23. M. Auer, C. Welter, and B. Kolb, *HS Application Report No. 114*, Bodenseewerk Perkin-Elmer Co., Überlingen, Germany, 1989.
24. U. Schwenk, H. Hachenberg, and M. Förderreuther, *Brennstoff-Chem.* **42**, 194–199, 295–296 (1962).
25. K. Porter and D. H. Volman, *Anal. Chem.* **34**, 748–749 (1962).
26. *Data Sheet No. GCHN-10*, Perkin-Elmer Corporation, Norwalk, CT, 1991.

6

Method Development
in HS-GC

The analyst is often faced with the problem of a new sample: how to prepare it, which method to use for quantitative analysis, and what conditions to select. It is practically impossible to give exact solutions before some preliminary investigations have been carried out. In this chapter, therefore, we present some general guidelines on how to proceed, along with three examples illustrating the various questions one must consider in method development before the final analytical method and conditions can be chosen. The following examples are discussed:

- determination of residual styrene monomer in polystyrene samples
- determination of residual solvents in a printed laminated plastic film
- determination of organic solvents in an electrolytic plating bath

We selected these examples because they represent complex, practical samples permitting a number of choices, hence requiring the analyst to make certain decisions. The examples discussed here also illustrate some of the quantitative methods of HS-GC treated in Chapter 5.

6.1 General Guidelines

Method development always starts with a general consideration of the sample. The principal questions to be answered are: Is the analyte to be determined volatile enough for HS analysis? Is the physical state of the sample suitable for placing a reproducible aliquot into the headspace vial? If the sample is a solid, will it represent a partition system or will it have to be modified to provide partitioning of the analyte between the two phases or its release into the headspace? Alternately, can the sample be dissolved?

The next step is to evaluate the time needed for equilibration. This can be done by placing a number of vials containing the same amount of the sample into the thermostat of the automated HS-GC instrument and thermostatting them for increasingly longer times using, for example, the progressive working mode. Plots similar to Figures 4-1 through 4-5 will tell the analyst whether equilibration can be achieved within a reasonable time at the selected temperature, and if this is the case, how much time is needed. The thermostatting temperature may be increased to reduce the equilibration time, but one must be careful not to use too high a temperature, since it might be detrimental to the sample and/or the pressure in the vial might be too high.

The concentration of the analyte must also be considered. If it is too high, the sample will have to be diluted. On the other hand, if it is too low, the headspace conditions (sample size, temperature) must be adjusted to increase the headspace sensitivity, and one may consider the use of cryogenic sample concentration. On the other hand, dilution may also be selected to eliminate the matrix effect.

Finally, one must also evaluate the chromatograms from the point of peak separation and adjust the chromatographic conditions, utilizing the general rules of gas chromatography. Investigation of the chromatogram can also provide information about whether there is a possibility of using an internal standard—in other words, whether there is an empty space for it in the chromatogram.

The conditions of the transfer of an aliquot from the headspace of the vials into the GC column also need special consideration. Since, however, these depend on the system used, no general guidelines can be given here: the analyst must consult the particular instrument's instruction manual and other documentation provided by the vendor.

When the general conditions are set, the next step is to select the method to be used for quantitative analysis. It is important to understand that the various quantitative techniques are usually interchangeable and the actual selection depends on a number of aspects: for example, whether it is a single analysis, versus a number of similar samples to be analyzed, and whether the aim is to analyze final products (where the time needed for the analysis is of secondary importance) or the measurements are related to production control (where the result is needed in a short time).

Multiple headspace extraction with external (vapor) standard is very useful as the first choice in method development because it gives a lot of information on the sample, its equilibration, and distribution. Based on the knowledge gained in MHE analysis, the headspace conditions can be further refined and adjusted. Also, based on the information obtained from the MHE measurement, one can then select another quantitative method for the routine analysis of the sample. Again, no general rules can be applied here: the sample and the requirements will ultimately determine the best method. In the following examples we try to illustrate the successive circumstances the analyst must consider before the final analytical method is established.

6.2 Determination of the Residual Monomer Content of Polystyrene Pellets

Our aim is to develop a method for the determination of the residual styrene monomer (SM) content of polystyrene (PS) pellets. For the analysis, the PS pellets were first freeze-ground with liquid nitrogen and the resulting powder was subjected to the investigations.

6.2.1 First Approach: Use of Internal Standard with MHE

We first selected the internal standard method with MHE measurement, using 2-methoxyethanol (methyl cellosolve; BP 124–125 °C) as the internal standard. Four-point MHE procedure was used in the determination of the response factor and a nine-point procedure in the analysis of the sample.

Example 6.1

Response factor determination. Into a 10 mL flask 1.0 mL (0.9074 g) of SM and 1.0 mL (0.9660 g) of methyl cellosolve were added and the flask was filled up with dimethylformamide (DMF). Into a headspace vial, 2.0 μL of this solution (containing 181.5 μg SM and 193.2 μg methyl cellosolve) was added and totally evaporated at 120 °C for 30 minutes. Analytical and instrumental conditions are given in the figure caption of Figure 6-1.

The results of the MHE measurements are listed in Table 6-1, together with the linear regression data; the corresponding plots are shown in Figure 6-2. The sums of the peak areas (total area ΣA_i and ΣA_{st} in Table 6-1) are calculated using eq. 5.28, and from these values the response factor (f_i or RF) is derived according to eq. 5.33:

$$f_i = \frac{181.5}{193.2} \cdot \frac{1{,}284{,}486}{3{,}897{,}588} = 0.3096$$

Sample analysis. A standard solution of methyl cellosolve was prepared by filling up 1.0 mL of the substance (0.9660 g) with DMF to 10.0 mL. A 2.5 μL aliquot of this solution (containing 241.5 μg of the standard) was added to 200 mg of the PS powder and equilibrated at 120 °C for 120 minutes. The results of MHE measurements are listed in Table 6-2, together with the regression data; the corresponding plots are shown in Figure 6-3.

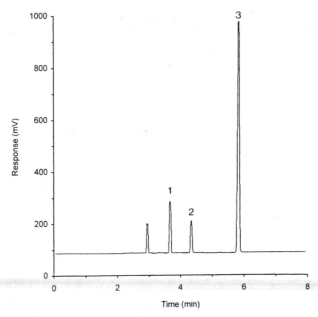

Figure 6-1. Analysis of residual styrene monomer in a polystyrene sample, using methyl cellosolve as the internal standard (Example 6.1). *HS conditions*: Sample: 200 mg freeze-ground polystyrene to which 241.5 μg of the internal standard was added in DMF solution. Equilibration at 120 °C for 2 hours. *GC conditions*: 50 m × 0.32 mm I.D. fused-silica, open-tubular column coated with bonded poly(ethylene glycol) stationary phase; film thickness: 0.4 μm. Column temperature: 120 °C. Split sampling. Flame-ionization detector. *Peaks*: *1* = methyl cellosolve, *2* = styrene, *3* = dimethylformamide.

Table 6-1   MHE measurement of residual styrene monomer in polystyrene pellets (Example 6.1); determination of the response factor f_i^*

| | Peak area (counts) for | |
| | Analyte (SM), A_i | Internal standard (methyl cellosolve), A_{st} |
i		
1	2,343,274	773,093
2	933,169	307,106
3	373,967	123,086
4	146,473	48,537
Linear regression		
Correlation coefficient r	−0.999987	−0.999995
Slope q	0.923139	0.921853
$Q = e^{-q}$	0.397270	0.397781
Intercept A_1^*	2,349,193	773,542
Total area ΣA	3,897,588	1,284,486

* For the linear regression plots, see Figure 6-2.

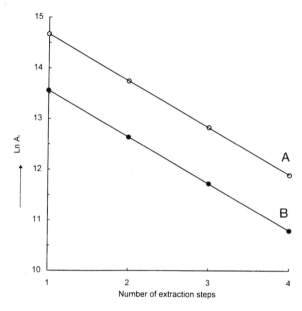

Figure 6-2. Linear regression plots for the response factor determination by MHE and TVT, outlined in Example No. 6.1. A = styrene monomer, B = methyl cellosolve.

Table 6-2 MHE measurement of residual styrene monomer in polystyrene pellets (Example 6.1); analysis by internal standard calibration

	Peak area (counts) for	
i	Analyte (SM), A_i	Internal standard (methyl cellosolve), A_{st}
1	478,194	756,587
2	371,329	398,658
3	276,909	202,251
4	209,592	104,783
5	154,916	53,510
6	116,022	28,129
7	85,186	14,364
8	64,049	7,590
9	47,010	3,873
Linear regression		
Correlation coefficient r	−0.999809	−0.999987
Slope q	0.29167	0.65979
$Q = e^{-q}$	0.74702	0.51696
Intercept A_1^*	493,159	759,434
Total area ΣA	1,949,378	1,572,191

* For the linear regression plots, see Figure 6-3.

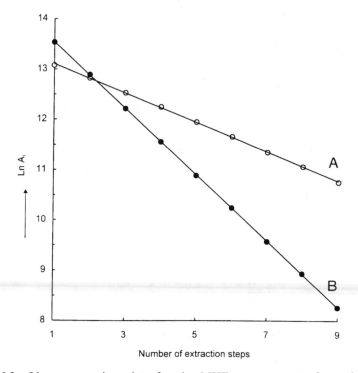

Figure 6-3. Linear regression plots for the MHE measurement of a polystyrene sample outlined in Example 6.1, using an internal standard. A = styrene monomer, B = internal standard (methyl cellosolve).

The amount of SM present can be calculated from the sums of the peak areas (total area ΣA_i and ΣA_{st} in Table 6-2) and the response factor f_i according to eq. 5.32:

$$W_i = 0.2415 \cdot 0.3096 \cdot \frac{1{,}949{,}361}{1{,}572{,}197} = 0.09271 \text{ mg}$$

The concentration in the sample is 92.7/0.2 = 464 μg/g.

6.2.2 Second Approach: Single Determination with Internal Standard

Based on the foregoing measurements we can now establish a simplified method for routine analysis. In this we would follow the procedure as given in Example 6.1 for "sample analysis," analyzing 200 mg of PS powder to which 2.5 μL of methyl cellosolve solution (containing 241.5 μg of the substance) was added. However, we shall carry out only a single HS measurement (see Figure 6-1). In this we obtain the peak area for the SM present in the analyzed sample $(A_{1,i})$, and for the added methyl cellosolve $(A_{1,st})$. For the calculation of the amount of SM present we now need a calibration factor f_c, which can be established from a single measurement of a solid sample in which the amount W of the standard st and the analyte i are known (cf. eq. 5.7):

$$f_c = \frac{W_i^c}{W_{st}^c} \cdot \frac{A_{st}^c}{A_i^c} \tag{6.1}$$

We are using superscript c here to indicate that these values reflect a calibration run.

The just analyzed sample can serve for this calculation as the calibration sample (or working standard): the amount of SM present was found to be $W_i^c = 92.7 \, \mu g$, while the amount of added methyl cellosolve was $W_{st}^c = 241.5 \, \mu g$. Using the peak areas obtained in the first measurement ($i = 1$) listed in Table 6-2,

$$A_{st}^c = 756,587$$

$$A_i^c = 478,194$$

the calibration factor calculated according to eq. 6.1 is $f_c = 0.6073$. Naturally, this value will differ from the value of the response factor determined in Example 6.1 ($f_i = 0.3096$), where the *response factor* was determined using fully evaporated substances and thus reflected only the differences in detector response. On the other hand, the *calibration factor* used here represents a combination of the response factor and the partition coefficients (matrix effects).

6.2.3 Third Approach: Use of External Standard with MHE

It is also possible to carry out the determination by MHE and using the external standard method. In Example 6.1 we have prepared a solution, with an SM concentration of $0.9074 \, g/10 \, mL$, and $2.0 \, \mu L$ of this solution containing $W_{ex} = 181.5 \, \mu g$ SM was used in the determination of the response factor. This solution can also be considered to be an external standard and A_i in Table 6.1 becomes now A_{ex}. MHE measurements of this solution resulted in $\Sigma A_{ex} = 3,897,588$ as the sum of the peak areas. Analyzing the polymer sample, $\Sigma A_i = 1,949,378$ (see Table 6.2) was obtained. Thus, (eq. 5.30):

$$W_i = \frac{\Sigma A_i}{\Sigma A_{ex}} \cdot W_{ex} = \frac{1,949,378}{3,897,588} \cdot 181.5 = 90.8 \, \mu g$$

corresponding to a concentration of $90.8/0.2 = 454.0 \, \mu g/g$, which differs only by 2% from the result obtained in Example 6.1.

Note that in these calculations no correction was made for the difference in the sample volume, because the volume correction factor is only 1.009 (i.e., the difference would be less than 1%).

6.2.4 Fourth Approach: Use of the Solution Approach

The measurement may also be carried out using the solution approach (see Section 4.2), dissolving the polystyrene sample in dimethylformamide and using an SM solution in DMF as the external standard. In this case, however, the dilution of the sample will result in a headspace sensitivity that is about one order of magnitude less.

6.3 Determination of Residual Solvents in a Printed Plastic Film

An important application of HS-GC is the determination of residual solvents present in printed, plastic, or aluminum films used in food packaging. Figure 6-4 shows a typical chromatogram of a sample. It was found that the slow diffusion in such a thick laminated film (250 μm) leads to a long equilibration time: 2 hours at 150 °C. A number of quantitative headspace techniques could be used for this determination, and the aim of method development is to evaluate the best possible method for routine analysis.

As shown in Figure 6-4, a number of residual solvents may be present; the examples discussed here refer to toluene only. The amount (concentration) of other solvents can be determined in a similar way, using the appropriate standards.

6.3.1 First Approach: Use of External Standard with MHE

Although it is too long for routine analysis, a multipoint MHE measurement with external standard calibration was selected as the first approach, using toluene vapor as the external standard. A 10 × 10 cm piece of the film was used as the sample: its volume is 2.50 mL, which is 11.2% of the vial's volume; thus, correction for the sample volume was necessary.

Example 6.2

A 10 cm × 10 cm printed, laminated plastic film was placed into a 22.3 mL headspace vial. Into a separate vial 1 μL (= 0.867 mg) of pure toluene was introduced. Both vials were thermostatted at 130 °C for 2 hours and then subjected to a multipoint MHE measurement. Columns A and B of Table 6-3 list the results, together with the regression data; the corresponding plots (A and B) are shown in Figure 6-5.

The volume correction factor is (cf. eq. 5.37):

$$f_V = \frac{22.3}{22.3 - 2.5} = 1.1263$$

The sums of the peak ares (total area ΣA in Table 6-3) are calculated using eq. 5.28. The value for the external standard is further corrected by the volume correction

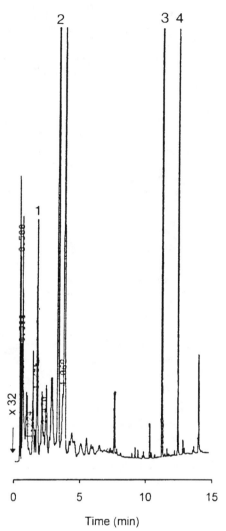

Figure 6-4. Analysis of residual solvents in a printed laminated plastic film (Example 6.2). *HS conditions*: Sample: 100 cm² × 0.25 mm thick film, equilibrated at 150 °C for 2 hours. *GC conditions*: 25 m × 0.32 mm I.D. fused-silica, open-tubular column coated with bonded poly(ethylene glycol) stationary phase; film thickness: 1 μm. Column temperature: isothermal at 60 °C for 5 minutes, then programmed at 10 °C/min to 160 °C. Splitless sampling. Flame-ionization detector. *Peaks*: *1* = diisobutyl ketone, *2* = toluene, *3* = butyl cellosolve acetate, *4* = hexyl cellosolve.

Table 6-3 MHE measurements related to the method development for the
determination of residual solvent (toluene) in a printed, laminated
plastic film (Examples 6.2 and 6.3)

	Peak area (counts) for		
	A	B	C
		External	Sample + 1 μL
i	Sample, A_o	standard, A_{ex}	toluene, $A_{(o+a)}$
1	51,540	475,983	264,986
2	27,270	131,875	169,109
3	15,473	40,600	98,626
4	8,736	12,296	56,766
5	5,183	3,672	33,978
6	3,269	1,083	19,605
7	1,939		12,006
8	1,235		6,753
Linear regression			
Correlation coefficient r	−0.998685	−0.999943	−0.999786
Slope q	0.53004	1.21046	0.52706
$Q = e^{-q}$	0.58858	0.29806	0.59034
Intercept A_1^*	46,402	460,789	277,105
Total area ΣA	112,785	656,451	676,424

* For the linear regression plots, see Figure 6-5.

factor and the corrected total area is obtained:

$$\Sigma A_{ex}^x = 656,451 \cdot 1.1263 = 739,361$$

From this and the sum of the peak areas of toluene from the sample (total area ΣA_o
in Table 6-3) the amount of toluene present in the sample is derived:

$$\frac{112,785}{739,361} \cdot 0.876 = 0.132\,\text{mg}$$

representing a concentration of 1.32 μg/cm^2.

6.3.2 Second Approach: Use of Standard Addition with MHE

The multipoint MHE approach is too time-consuming for routine measurements, particularly if the equilibration time is long, because the overlapping
working mode is not applicable here. The next step is to ascertain whether
standard addition can be used. Here we first have to test whether gas phase
addition (GPA) works with this fairly thick sample: in other words, whether
the amount of toluene that is added to the gas phase of the vial will partition

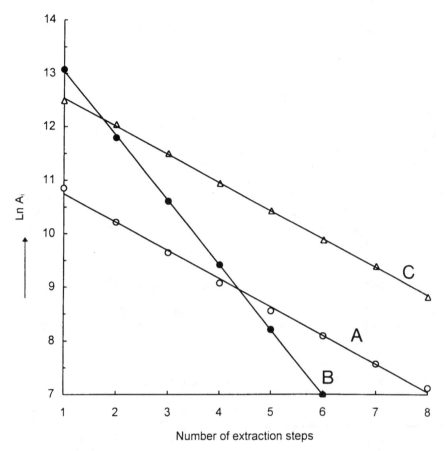

Figure 6-5. Linear regression plots for the determination of residual toluene in a printed plastic film. $A = 100 \, cm^2$ film, B = external calibration standard, $C = 100 \, cm^2$ film + 1 μL toluene. For details see Examples 6.2 and 6.3.

sufficiently fast between the two phases. Again, MHE measurement can be used for this test: as we have seen (case E in Figure 5-15), if the two plots are parallel (having the same slope value), an identical matrix effect (i.e., the same partitioning process) is indicated.

Example 6.3

A 10 cm × 10 cm printed, laminated plastic film was placed into a headspace vial and 1 μL of toluene (=0.867 mg) was added to it. The sample was carried through the same equilibration and MHE procedure as the earlier samples (Example 6.2). The results, together with the regression data, are listed in column C in Table 6-3 (plot C in Figure 6-5). From these data, the sum of the peak areas (total area $\Sigma A_{(o+a)}$) is calculated using eq. 5.28.

The amount of toluene present is calculated relative to the analysis of the first sample (column A in Table 6-3), according to eq. 5.34:

$$W_o = 0.867 \cdot \frac{112,785}{676,424 - 112,785} = 0.173 \text{ mg}$$

and its concentration is $1.73 \text{ } \mu g/cm^2$.

Comparing plots A and C in Figure 6-5 we can see that they are parallel, and Table 6-3 shows that their slopes are practically identical. This is important information, indicating that we have indeed a partition system and the standard added to the gas phase freely equilibrates between the two phases.

However, we have another problem here: the discrepancy of the results, compared to the results of Example 6.2 (1.32 vs. $1.73 \text{ } \mu g/cm^2$). The reason for this difference is the inherent sample homogeneity, which is commonly found in this type of sample. There are two ways this problem may be overcome; these are indicated in the third and fourth approaches.

6.3.3 Third Approach: Use of Internal Standard

The best solution would be to use an internal standard. However, in this particular case, it could not be applied. This is clear if we look at the chromatogram in Figure 6-4: there is simply no space for an additional peak.*

6.3.4 Fourth Approach: Use of Standard Addition

The final approach we selected was standard addition, as used in the second approach, but (1) compensating for sample inhomogeneity by the use of an internal normalization standard and (2) using only a single measurement to reduce the time needed for the analysis. We used peak 1 in the chromatogram (Figure 6-4), corresponding to diisobutyl ketone, as the internal normalization standard: now the calculation can be carried out using eq. 5.19.

Example 6.4

The following peak area values were obtained:

	Peak area (counts)	
	Toluene	Normalization standard
Original sample	51,540	9,120
Sample + 1 μL toluene	264,986	6,473

---

* Incidentally, this is a common problem associated with the use of internal standards in a complex sample; it is not specific to headspace analysis.

The values for the normalization standard are (cf. eqs. 5.16-5.17):

$R_o = 51,540/9120 = 5.651$

$R_{(o+a)} = 264,986/6473 = 40.937$

The amount of toluene present in the original sample is (eq. 5.19):

$$W_o = 0.867 \cdot \frac{5.651}{40.9037 - 5.651} = 0.139 \text{ mg}$$

and its concentration is $1.39 \, \mu g/cm^2$ which agrees well with the original measurement given in Example 6.2 ($1.32 \, \mu g/cm^2$).

6.4 Determination of the Volatile Constituents of a Cathodic Electrolytic Plating Bath

In this example we first tried to eliminate the matrix effects without dilution, by using the full evaporation technique — which seemed to be the logical solution. However, the results were unsatisfactory, and therefore another way had to be found.

The sample consisted of a cathodic electrolytic plating bath, and we were looking for the concentration of ethylene glycol monohexyl ether (hexyl cellosolve: BP = 208 °C) in it. In the present case we felt that the use of FET with an undiluted $10 \, \mu L$ sample would simplify the determination without the necessity of reducing headspace sensitivity by dilution. Using 130 °C and 30 minutes for thermostatting, we assumed that all the hexyl cellosolve would evaporate.

6.4.1 First Approach: Use of External Standard with MHE

As the first approach we injected $10 \, \mu L$ of an aqueous solution of hexyl cellosolve as the external standard into a headspace vial and the same volume of the plating bath sample into another one. Both samples were carried through an eight-step MHE procedure, and the corresponding plots are given in Figure 6-6.

If we look at the plot related to the external standard, reasonably good linearity can be found. This means that the conditions were satisfactory for full evaporation. However the electrolytic bath shows nonlinear results: the plot connecting the individual points for hexyl cellosolve corresponds to case F in Figure 5-15, which was identified in Chapter 5 as being due to additional adsorption effects of the various solid compounds (salts, pigments, and other ingredients) in the plating bath which remained as a dry and highly adsorptive residue in the vial. This indicated that FET cannot be used in this case.

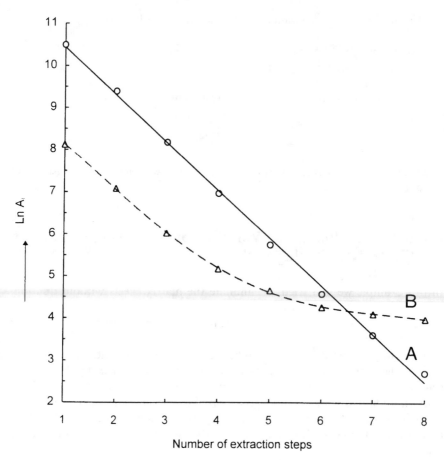

Figure 6-6.   Linear regression plots for the MHE determination of hexyl cellosolve in a cathodic electrolytic plating bath, thermostatted at 130 °C for 30 minutes. A = aqueous solution of hexyl cellosolve, B = plating bath.

6.4.2 Second Approach: Dilution and Use of External Standard

However, the volatile sample components are present in sufficiently high concentrations, and this allowed us to eliminate any matrix influence by diluting the plating bath 1:10 with water. Figure 6-7 shows the chromatogram obtained from a 2 mL aliquot. For quantitative evaluation, the aqueous solution of hexyl cellosolve was used here as the external standard. This procedure gave satisfactory results and was used to further process a series of similar samples (one was shown earlier in Figure 5-7).

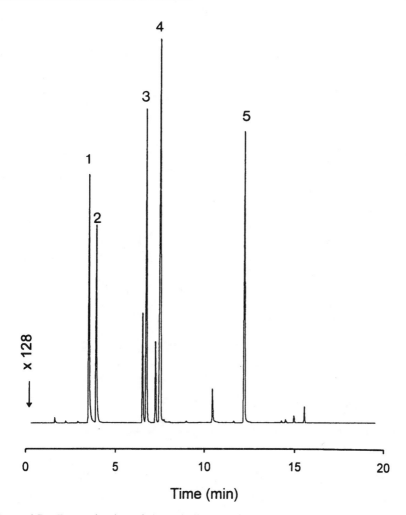

Figure 6-7. Determination of the volatile organic constituents present in a cathodic electrolytic plating bath. *HS conditions*: Original sample diluted 1:10 with water; 2 mL of the diluted solution analyzed. Equilibration at 90 °C for 60 minutes. *GC conditions*: 25 m × 0.25 mm I.D. fused-silica, open-tubular column coated with bonded methyl silicone stationary phase; film thickness: 1 μm. Column temperature: 2 minutes isothermal at 50 °C, then programmed at 8 °C/min to 180 °C. Split sampling. Flame-ionization detector with attenuation of × 128. *Peaks* (concentration in the original sample): *1* = ethyl cellosolve (1.0%), *2* = methyl isobutyl ketone (0.03%), *3* = xylene (0.08%), *4* = butyl cellosolve (0.9%), *5* = hexyl cellosolve (0.9%).

CHAPTER

7

Nonequilibrium Static Headspace Analysis

In all the preceding discussions we assumed that an equilibrium was established between the sample and its headspace prior to the transfer of an aliquot of the headspace into the GC column. In such a case the theoretical relationships discussed in Chapter 2 are valid and the concentration (amount) of the analyte in the original sample can be established from the quantitative evaluation of the results of the aliquot's analysis.

Sometimes, however, one may carry out headspace analysis before equilibrium is reached and use these data in further investigations, including quantitative analysis. Naturally, the prerequisite for this is perfect reproducibility of all operating and analytical parameters.

Headspace analysis may be carried out under nonequilibrium conditions for two reasons. In the first case, the analysis time — particularly the time needed for equilibration — would be too long for the intended purpose of process control or for routine measurements. Therefore, the sample is thermostatted for less time than is needed for equilibrium, and quantitative results are obtained with help of predetermined calibration factors. In the second case the sample is heat sensitive and therefore it might be damaged in the course of full equilibration.

7.1 Accelerated Analysis

In the first case the necessary equilibration time and the correct quantitative results are known. If, however, this time is considered to be too long because a fast result is required (e.g., for process control or in the case of a priority

Table 7-1 Determination of the menthol
content of scented cigarette tobacco
by standard addition
(Example 7.1)*

Amount of menthol added (μg)	Peak area (counts)[†]
0	2359
500	3126
1000	4049
1500	4648
Linear regression	
Correlation coefficient r	0.99695
Slope a	1.5580
Intercept b	2377

* For the regression plot, see Figure 7-1.
† Averages of three measurements, with a mean relative standard deviation of 2.85%.

sample), the thermostatting time may be shortened and the analytical results — which will be too low, since equilibrium has not been reached — adjusted to obtain the correct concentration value. This possibility is illustrated here with respect to the determination of menthol in tobacco.

For these measurements, aliquots of the tobacco were weighed into headspace vials and water was added as a displacer. The time needed for full equilibration was found to be surprisingly long — 90 minutes at 80 °C — and this is apparently caused by the slow desorption process in the tobacco matrix rather than by the partitioning process of menthol between the aqueous phase and the gas phase, which is much faster. The measurement can be conveniently carried out by standard addition as outlined in Example 7.1. Evaluation was carried out using linear regression according to eq. 5.20.

Example 7.1

Scented tobacco samples (500 mg) were placed into four headspace vials and 5 mL water was added to each. Separately a methanolic solution of menthol, with a concentration of 50 mg/mL, was prepared, and 10, 20, and 30 μL aliquots of this solution were added to vials 2, 3, and 4. The vials were thermostatted at 80 °C for 90 minutes and then their headspace was analyzed in the usual way, on a 50 m × 0.32 mm I.D. fused-silica, open-tubular column coated with bonded poly(ethylene glycol) stationary phase (film thickness: 0.4 μm). The column was kept isothermally at 130 °C for 5 minutes and then programmed at 6 °C/min; the retention time of menthol was between 6 and 7 minutes. Table 7-1 lists the peak area values together with the regression data; the regression plot is given in Figure 7-1. The calculated amount of menthol in the sample is (cf. eq. 5.21) $b/a = 1525.7\,\mu$g. The concentration of menthol in the tobacco sample is 0.31 wt%.

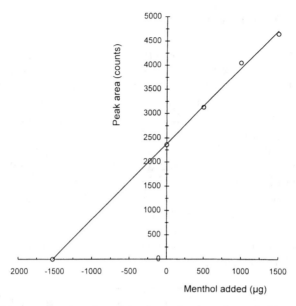

Figure 7-1. Linear regression plot of the determination of menthol in scented tobacco by standard addition. For details, see Example 7.1.

Although this determination takes too long for the intended purpose of process control, the plot of peak area versus thermostatting time (Figure 7-2) shows that 75% of the final peak area is reached in 15 minutes. Thus, in routine analysis, the sample is thermostatted at 80 °C for only 15 minutes, and then the result is multiplied by a factor of $100/75 = 1.33$.

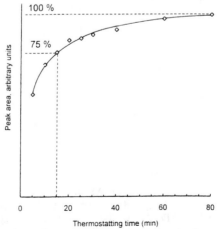

Figure 7-2. Equilibration of a scented tobacco sample for menthol. Temperature: 80 °C.

7.2 Heat-Sensitive Samples

The second reason for selecting a short thermostatting time (i.e., less than needed for equilibration) is connected with heat-sensitive samples, which might be altered during a prolonged thermstatting time at an elevated temperature. If a short thermostatting time at a moderate temperature, is used, however, HS-GC may still give valuable information on the relative amounts of volatile compounds present.*

Such investigations were described by Shinohara et al. [1], studying the possibility of using HS-GC for the characterization of the flavor of fresh vegetables. These investigators placed 1 g of chopped vegetables (or juices prepared from them) into the headspace vials and thermostatted the sample at 90 °C for 10 minutes (which is too short for equilibrium). The chromatograms obtained were evaluated both qualitatively, using a mass spectrometer directly coupled to the outlet of the separation column for compound identification, and quantitatively, establishing the relative peak area values. Two examples from this work are shown here.

---

* Another possibility for heat-sensitive samples was already discussed in Section 3.8, which explained the use of a cryofocusing enrichment technique to compensate for the reduced sensitivity if the sample is analyzed under safe conditions at room temperature.

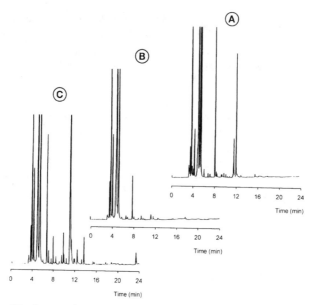

Figure 7-3. Headspace chromatograms of (A) raw tomato, (B) tomato juice, and (C) vegetable juice . *HS conditions*: Sample: 1 g; thermostatted at 90 °C for 10 minutes. *GC conditions*: 50 m × 0.25 mm I.D. fused-silica, open-tubular column, coated with poly(ethylene glycol) 20M stationary phase; film thickness: 0.3 μm. Column temperature: programmed from 50 °C to 210 °C at 4 °C/min. Split sampling (1:15). Flame-ionization detector.

Source: Reproduced from ref. 1 with permission of the authors and *Chromatographia*.

Figure 7-3 shows comparative chromatograms obtained from raw tomato, tomato juice, and a vegetable juice of which tomato was the main component. The tomato juice chromatogram contains the early peaks of the raw tomato chromatogram, while — as expected — the vegetable juice chromatogram is much more complex.

Figure 7-4 shows the duplicate analysis of raw parsely. The corresponding computer printout indicated the presence of a total of 57 peaks. Portions of this full evaluation sheet are given in Table 7-2, which presents data for 18 peaks (representing 92% of the total), indicating retention time, peak area, and area percent values, and identifying the most important peaks. These data demonstrate the remarkable reproducibility of HS-GC analysis. The differences in the retention times are only in the third decimal place, representing less than 0.5 second; the individual *absolute* peak area were reproduced within ±3%, except for the smallest peak (46), corresponding to only 0.007% of the total peak area. We emphasize that the peaks represent a range of 1:4071, from 0.007% to 28.5%. How well the absolute values were reproduced is best

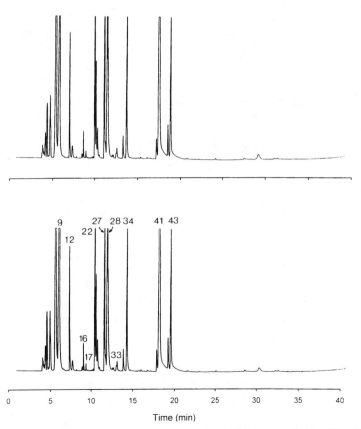

Figure 7-4. Duplicate analysis of raw parsley by HS-GC. For peak identification and quantitative data see Table 7-2. *HS and GC conditions*: see Figure 7-3.

Source: Reproduced from ref. 1 with permission of the authors and *Chromatographia*.

Table 7-2. Duplicate analysis of 1.00 g of raw parsley: Equilibration at 90 °C for 10 minutes (see Figure 7-4).

Peak no.	Identified compound	First run			Second run			Mean values			
		Retention time (min)	Peak area (counts)	Peak area (%)	Retention time (min)	Peak area (counts)	Peak area (%)	Retention time (min)	(±%)	Peak area (counts)	(±%)
03		3.401	810	0.053	3.401	994	0.066	3.4010	0.000	902	0.200
08	α-Pinene	4.810	244,131	15.988	4.811	239,324	15.990	4.8105	0.010	241,727	0.994
09	Camphene	5.226	252,492	16.545	5.231	245,847	16.424	5.2285	0.048	249,169	1.333
12	β-Pinene	6.520	30,157	1.976	6.514	29,961	2.002	6.5170	0.046	30,059	0.326
16	Sabinene	8.270	5,293	0.347	8.263	5,007	0.335	8.2665	0.042	5,150	2.777
17		8.579	1,539	0.101	8.572	1,533	0.102	8.5755	0.041	1,536	0.195
22	Myrcene	9.622	68,176	4.468	9.616	64,092	4.282	9.6190	0.031	66,134	3.088
27	α-Phellandrene	10.768	64,179	4.205	10.762	64,101	4.283	10.7650	0.028	64,140	0.061
28	β-Phellandrene	11.105	189,075	12.391	11.100	188,827	12.616	11.1025	0.023	188,951	0.066
33	p-Cymene	13.040	6,791	0.445	13.034	6,501	0.434	13.0370	0.023	6,646	2.182
34	Terpinolene	13.434	47,762	3.130	13.526	45,607	3.047	13.4800	0.341	46,684	2.309
37		15.087	437	0.029	15.081	461	0.031	15.0840	0.020	449	2.673
41	1,3,8-p-Menthat-riene	17.258	434,056	28.445	17.255	426,546	28.398	17.2565	0.009	430,301	0.873
43	Isopropenyltoluene	18.612	53,362	3.497	18.609	52,287	3.493	18.6105	0.008	52,824	1.018
46		23.802	104	0.007	23.797	122	0.008	23.7995	0.011	113	7.965
48		26.451	179	0.012	26.480	173	0.012	26.4655	0.055	176	1.705
53		29.319	6,383	0.418	29.316	6,605	0.441	29.3175	0.005	6,494	1.709
56		31.605	430	0.028	31.613	445	0.030	31.6090	0.013	437	1.714
Sum of 18 peaks			1,405,266	92.091		1,378,433	92.095			1,391,849	0.964
Sum of 57 peaks			1,525,955	100.000		1,496,756	100.000	100.000		1,511,355	0.966

Source: Data from Shinohara et al. [1]; the original printout contained 57 peaks; here a selected listing of 18 peaks is given.

demonstrated by pointing out that the sums of the absolute peak area (57 peaks!) were within $\pm 0.97\%$

These examples show that with the automated HS-GC systems, nonequilibrium static headspace analysis can also provide highly reproducible data and valuable information.

Reference

1. A. Shinohara, A. Sato, H. Ishii, and N. Onda, *Chromatographia*, **32**, 357–364 (1991).

8

Qualitative Analysis by HS-GC

In Chapter 5 we discussed in detail the methods of quantitative analysis by headspace–gas chromatography. However, one does not always need quantitative information: in many cases a qualitative evaluation of the chromatogram is sufficient. This can, for example, demonstrate changes in one sample relative to another, serve to identify a sample by pattern recognition using the headspace chromatogram as a "fingerprint," or facilitate the investigation of volatile components present in a complex sample that otherwise would need a complicated pretreatment procedure prior to analysis.

With respect to the use of HS-GC for the qualitative indication of changes in a sample, Figure 8-1 shows an example from an early use of headspace–gas chromatography in pollution control [1]. Chromatogram A was obtained from the HS-GC analysis of an untreated sewage sludge in a water purification plant, while chromatogram B was from a water sample taken after purification, demonstrating the effect of sewage treatment and the cleanness of the effluent from the purification plant. A similar example for qualitative changes, now from the food field, is given in Figure 8-2 [2]. Here chromatogram A was obtained by the HS-GC analysis of roasted (at 175 °C) and ground coffee beans. An aliquot of the coffee powder was then used to brew a pot of coffee and the remaining grounds were air-dried and analyzed by HS-GS under the same conditions as the fresh coffee powder. Comparison of the two chromatograms shows that all the volatile compounds present in the fresh coffee powder (which were formed during roasting of the coffee beans) disappeared: these were extracted by the water in the coffee machine during brewing and became constituents of the coffee drink.

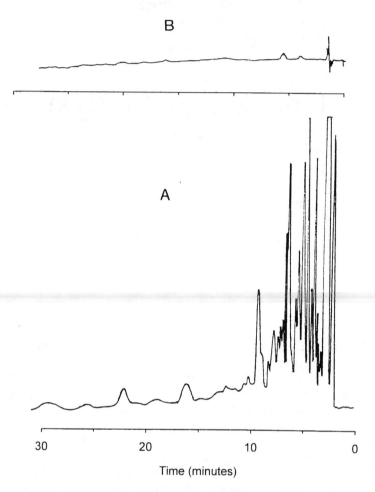

Figure 8-1. Headspace chromatograms: A = a raw sewage sludge, B = the water effluent of the purification plant, after treatment [1]. *HS conditions*: Sample: 1 mL, equilibration at 50 °C. *GC conditions*: Column: 2 m × 1/8 in. O.D. packed, containing 15% Carbowax 1500 stationary phase on Celite 60/80 mesh. Column temperature: 80 °C, isothermal. Flame-ionization detector.

Without headspace sampling, the use of a fairly complicated sample pre-treatment would have been needed in both cases, involving extraction and concentration of the obtained solutions. By using HS-GC, a representative sample of the volatile sample components could be directly transferred into the GC column. In this way the sample remained intact, nonvolatile material did not contaminate the GC system and, what is particularly important, no solvent peak interfered with the sample peaks. For example, it has been noted [3] with respect to the extraction of biological samples with diethyl ether that in the

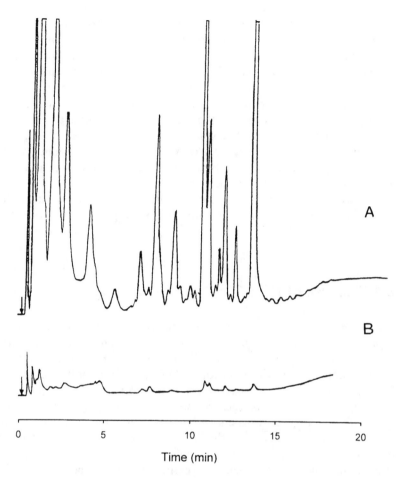

Figure 8-2. Analysis of roasted ground coffee (*A*) and grounds (*B*), after extraction in the coffee machine [2]. *HS conditions*: Sample: 0.7 g, equilibrated at 140 °C. *GC conditions*: Column: 6 ft × 1/8 in. O.D. packed, containing 8% SP-1000 on GasChrom Q, 100/120 mesh. Column temperature: 1 minute isothermal at 40 °C, then programmed at 10 °C/min to 220 °C. Carrier gas: Helium, 40 mL/min. Flame-ionization detector.

subsequent HS-GC analysis of the extract, the ether peak occasionally obscures some early metabolite peaks. This would be true for any investigation involving solvent extraction, and any impurities present in the solvent would also interfere with the chromatogram, which is a particular problem in trace analysis.

One remark is in order here. Obviously such comparative chromatograms include only the *volatile* compounds present and do not provide a complete picture of sample composition, whereas when extraction is used, all the soluble sample constituents are removed, including also those having no appreciable vapor pressure at the equilibration temperature, hence not present in the vial's headspace at detectable levels even though they would be included in the chromatogram of the extract. On the other hand, if the sample contains highly volatile components at low concentrations, these will be enhanced in the headspace, and thus their peaks will be prominent in the headspace chromatogram, while the corresponding peaks in the chromatogram of the extract will be relatively much smaller: in other words, headspace analysis actually enhances the peaks of volatile trace components. Thus, the chromatograms of the sample headspace and extract will be quite different. This is similar to a situation demonstrated earlier: Figure 2-3 compared the headspace chromatogram of a complex liquid sample (a wide-boiling-range hydrocarbon mixture) with the direct chromatogram of the original liquid sample.

8.1 The Use of HS-GC in "Fingerprinting"

It was recognized early in the development of gas chromatography that the profiles of the volatile compounds present in a natural substance are highly characteristic and may serve as "fingerprints" for comparison with other similar samples, even without a complete identification of the individual peaks. The direct analysis of the headspace of a complex sample has long provided a relatively easy method for such studies. The pioneering work of Mackay [4] Teranishi and their co-workers [5] in 1961 showed that even with the low-resolution packed columns of that time, meaningful chromatograms could be obtained for comparison. For example, Teranishi and Buttery [5] demonstrated the difference between the volatiles present in pears of two different types, and Mackay et al. [4] showed pattern differences in good versus deteriorated peppermint oil, or between natural banana and an imitation banana flavor: they called the use of static HS-GC an "objective measurement of odor."

In the past 30 years column efficiency and the technique itself have much improved. The potentials of static HS-GC, combined with cryogenic sampling and high resolution open-tubular columns, are illustrated here by two examples, demonstrating the different patterns of the volatile compounds causing the odor of two flowers: lily of the valley (*Convalleria majalis*) (Figure 8-3) and wisteria (*Wisteria sinensis*) (Figure 8-4). In both cases cryogenic sample concentration with 60-second sample transfer time was used. A similar example was shown earlier in Figure 3-28 for the HS-GC analysis of fennel seeds, equilibrated under safe conditions at room temperature, with cryogenic trapping and a transfer time as long as 9.9 minutes.

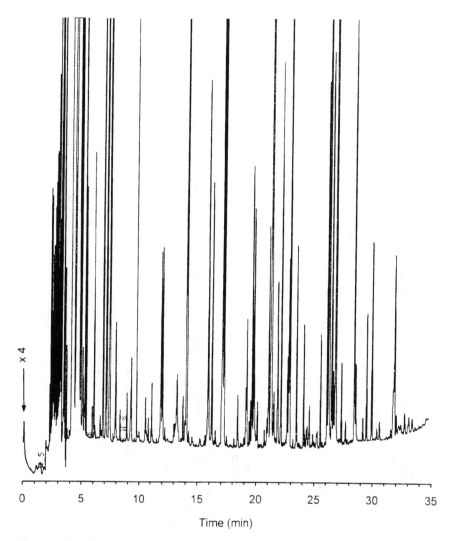

Figure 8-3. Chromatogram of the headspace of lily of the valley (*Convallaria majalis*) flowers, after cryofocusing. *HS conditions*: Sample: Eight flower heads, equilibrated at 80 °C for 30 minutes. Headspace transfer time: 60 seconds. Cryotrap and transfer line: 1 m × 0.32 mm I.D. fused-silica, open-tubular column coated with bonded methyl silicone stationary phase; film thickness: 5 μm. *GC conditions*: Column: 50 m × 0.25 mm I.D. fused-silica, open-tubular, coated with bonded phenyl (5%) methyl silicone stationary phase; film thickness: 1 μm. Column temperature: 8 minutes isothermal at 45 °C, then programmed at 8 °C/min to 120 °C, and from there at 6 °C/min to 250 °C. Splitless injection. Carrier gas: Hydrogen. Flame-ionization detector at attenuation of × 4.

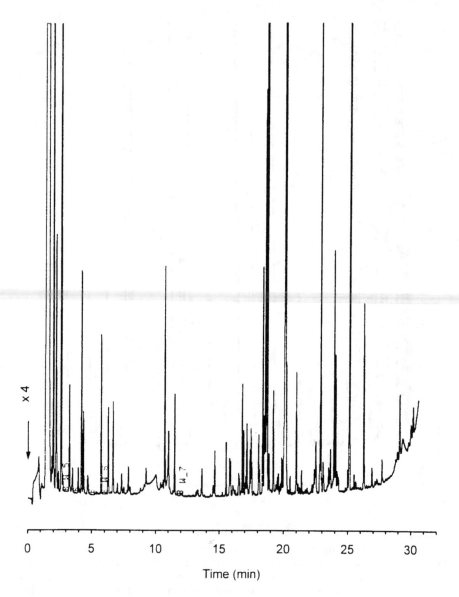

Figure 8-4. Chromatogram of the headspace of wisteria (*Wisteria sinensis*) flowers, after cryofocusing. For HS and GC conditions, see the caption of 8-3.

8.2 The Use of Headspace Sampling in Hyphenated Systems

A very important advantage of gas chromatography is the possibility of its direct coupling with some ultimate identification systems such as a mass spectrometer (MS) or a Fourier-transform infrared spectrophotometer (FTIR) (see, e.g., ref. 6). Such directly coupled, combined instruments are characterized today as *hyphenated systems*. Here, again, the possibility of direct sampling of the volatile sample components without any pretreatment represents a great advantage.

Chapter 7 already presented some examples for the direct combination of headspace sampling, gas chromatography, and mass spectrometry. For the identification of the volatile compounds present in raw vegetables by any other method, a fairly complicated sample treatment procedure would have been needed. When static headspace sampling of the raw vegetable was used, however, aliquots of the volatile sample components could be directly introduced into the GC column and identified by mass spectrometry. The comparison of tomato and vegetables juices with raw tomato (Figure 7-3) showed how various relatively similar samples can be characterized, while the investigation of raw parsley (Figure 7-4) demonstrated the ease of direct GC-MS identification of the volatile compounds after headspace sampling and the excellent reproducibility of the system. Figure 8-5 shows one more example from the same work [7]: the identification of the most characteristic volatile substances, mostly alkyl disulfides, present in raw onion, by the mass spectrometric investigations of the column effluent.

A very important application of GC-MS is found in environmental analysis, in the identification of volatile halogenated and aromatic hydrocarbons in drinking water. Here, static headspace sampling provides an excellent way to achieve high sensitivity, and the HS-GC systems can be directly coupled to a mass spectrometer. An application of a complex HS-GC-MS system for water analysis is demonstrated in Figure 8-6 [8]: here, the water standard analyzed contained 44 compounds at the 10 μg/L (ppb) level, and each was identified by its mass spectrum. For peak identification, see Table 8-1.

A very interesting possibility for qualitative analysis is the combination of static headspace–gas chromatography with Fourier-transform infrared spectroscopy (FTIR). In GC-FTIR systems the chromatograms can be obtained in two ways: by the so-called Gram–Schmidt reconstruction, or by having a flame-ionization detector either parallel to or after the light pipe of the spectrophotometer. For details of the systems and the recommended conditions, see ref. 9. Due to the rather low sensitivity of FTIR, compared with other GC detectors, enrichment by cryofocusing is particularly useful. The Gram–Schmidt chromatogram in Figure 8-7 of the aroma components from hop pellets shows such a combination with 3 minutes transfer time onto a 0.32 mm I.D. fused-silica, open-tubular column [10]; for peak identification, see Table 8-2. Since water was not removed by a water trap (see Section 3.8.3), the

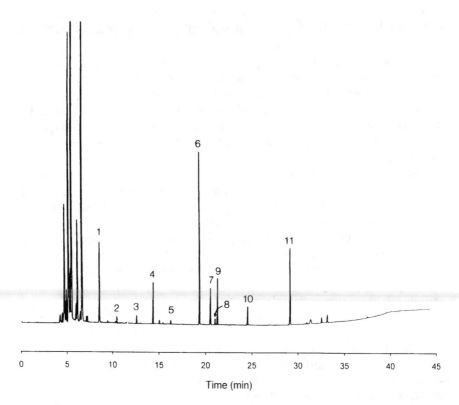

Figure 8-5. Chromatogram of the headspace of raw onion. *HS conditions*: Sample: 1 g, thermostatted for 10 minutes at 90 °C. Sample transfer time: 9 seconds. *GC conditions*: Column: 50 m × 0.25 mm I.D. fused-silica, open-tubular, coated with Carbowax 20M poly(ethylene glycol) stationary phase; film thickness: 0.3 μm. Column temperature programmed, from 50 °C to 210 °C at 4 °C/min. Split sampling, 1:15 split. Flame-ionization detector. Carrier gas: Helium, average velocity: 44 cm/s. *Peaks* (identified by MS): *1* = hexanal, *2* = 2-methyl-pentanal, *3* = methyl propyl disulfide, *4* = 2,4-di-methylthiophene, *5* = *trans*-methyl 1-propenyl disulfide, *6* = dipropyl disulfide, *7* = *cis*-1-propenylpropyl disulfide, *8* = allylpropyl disulfide, *9* = *trans*-1-propenylpropyl disulf-ide, *10* = propionic acid, *11* = dipropyl trisulfide.

Source: Reproduced from ref. 7 with permission of the authors and *Chromatographia*.

chromatogram shows the asymmetrical water peak. The identified sample constituent 2-methyl-3-buten-2-ol (peak 7) was found to be the pharmacologically active compound, responsible for the sedative effect of hops.

McClure has investigated by HS-GC-FTIR combination the components of various sassafras products [11,12]. The leaves of sassafras (*Sassafras albidum*) are an important ingredient in Louisiana Cajun cuisine, in the preparation of soups and gumbo, while root bark oil of sassafras trees had been used for some time in the preparation of "root beer"-type carbonated beverages. Around 1960 it was reported that safrole is carcinogenic, and therefore the use of sassafras

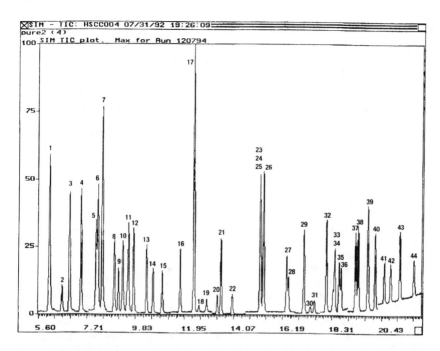

Figure 8-6. Single-ion monitored total-ion chromatogram of a water standard containing halogenated and aromatic hydrocarbons at the 10 ppb level, by directly coupled HS-GC-(quadrupole) MS. For peak identification see Table 8-1. *HS conditions*: Sample: 5 mL of the standard, equilibrated at 40 °C for 30 minutes. Sample transfer time: 4.8 seconds. *GC conditions*: Column: 60 m × 0.32 mm I.D. open-tubular, coated with VOCOL stationary phase (Supelco, Inc.); film thickness: 3 μm. Column temperature: 5 minutes isothermal at 40 °C, then programmed at 25 °C/min to 100 °C, and then at 5 °C/min to 180 °C.
Source: Reproduced from ref. 8 with permission of the author.

root bark oil was banned for use in root beer. McClure succeeded in determining safrole in sassafras root bark, but not in the cooking products of sassafras leaves at a detectable level.*

8.3 The Use of HS-GC in Microbiology

HS-GC has great potential in the investigation of the metabolic products in biological materials. The highly characteristic profiles of these compounds may serve as "fingerprints" for comparison with other similar samples, or with a "normal" sample, even without an identification of the individual peaks.

---

* Most recent measurements using supercritical–fluid extraction and subsequent GC-MS measurements (utilizing the higher sensitivity of MS as compared to FTIR) indicate that trace amounts of safrole may be present in the sassafras leaves [13].

Table 8-1   Identifcation of the peaks in Figure 8-6

Peak no.	Compound	Peak no.	Compound
1	1,1-Dichloroethylene	23	Chlorobenzene
2	Dichloromethane	24	Ethylbenzene
3	trans-1,2-Dichloroethylene	25	1,1,1,2-Tetrachloroethane
4	1,1-Dichloroethane	26	m- + p-Xylene
5	2,2-Dichloropropane	27	o-Xylene
6	cis-1,2-Dichloroethylene	28	Styrene
7	Chloroform	29	Isopropylbenzene
8	1,1,1-Trichloroethane	30	Bromoform
9	1,1-Dichloropropane	31	1,1,2,2-Tetrachloroethane
10	Carbon tetrachloride	32	n-Propylbenzene
11	1,2-Dichloroethane	33	Bromobenzene
12	Benzene	34	1,3,5-Trimethylbenzene
13	Trichloroethylene	35	o-Chlorotoluene
14	1,2-Dichloropropane	36	p-Chlorotoluene
15	Bromodichloromethane	37	tert-Butylbenzene
16	cis-1,3-Dichloropropylene	38	1,2,4-Trichlorobenzene
17	Toluene	39	sec-Butylbenzene
18	trans-1,2-Dichloropropylene	40	p-Isopropyltoluene
19	1,1,2-Trichloroethane	41	m-Dichlorobenzene
20	1,3-Dichloropropane	42	p-Dichlorobenzene
21	Tetrachloroethylene	43	n-Butylbenzene
22	Dibromochloromethane	44	o-Dichlorobenzene

A particular field in which the usefulness of static HS-GC for "fingerprint-ing" of biological material has been demonstrated is the identification of anaerobic bacteria by the determination of the volatile metabolic fermentation products such as volatile fatty acids, alcohols, and carbonyl compounds formed in liquid growth media. According to the classical methods [14,15], the media were extracted (e.g., with ether), the extract was subjected to gas chromato-graphic analysis, and the bacteria were identified by pattern recognition of the resulting chromatograms: many anaerobic bacteria generated "fingerprints" that are specific on a genus or species level.

Static HS-GC represents a favorable alternative to the traditional ether extraction, since sample preparation is considerably simplified and lends itself to automation. The latter consideration is particularly important in this field due to the need to investigate a large number of samples [3]. As an illustration, Figure 8-8 compares the headspace chromatograms of two similar bacteria, grown in the same medium; the striking difference in the patterns is evident.

In such investigations one is primarily looking for the pattern of the volatiles. Since, however, the number of volatile compounds present is limited, the result obtained from a culture in a certain media may not permit unambiguous identification. Therefore, Seifert and co-workers proposed that

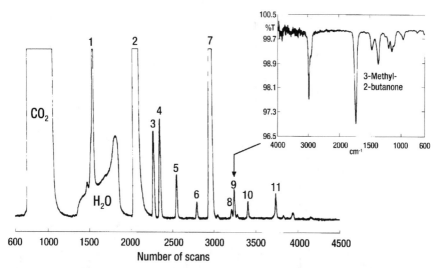

Figure 8-7. Chromatogram from aged hop obtained by direct coupling of HS-GC-FTIR with cryofocusing. *HS conditions*: Sample: 1 g hop pellets, thermostatted for 30 minutes at 80 °C. *GC conditions*: Column: 50 m × 0.32 mm I.D. fused-silica, open-tubular, coated with bonded methyl silicone stationary phase; film thickness: 5 μm. Column temperature: 36 °C isothermal for 3 minutes, then programmed at 5 °C/min to 90 °C, and then at 10 °C/min to 280 °C. For peak identification see Table 8-2. The FTIR spectrum is from peak 9.

Source: Reproduced from ref. 10 with permission of the authors and *Chromatographia*.

Table 8-2 Identification of the peaks in Figure 8-7

Peak no.	Compound
1	Methanol
2	Acetone
3	2-Methyl-1,3-butadiene
4	Methyl acetate
5	Isobutyric aldehyde
6	2-Butanone
7	2-Methyl-3-butene-2-ol
8	Isovaleric ldehyde
9	3-Methyl-2-butanone*
10	Methyl isobutyrate
11	4-Methyl-2-pentanone

* FTIR spectrum shown in the inset of Figure 8-7.

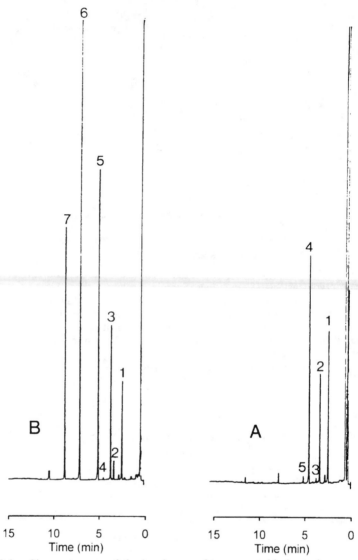

Figure 8-8. Chromatogram of the headspace of two anaerobic bacteria, grown in an aqueous PYG medium: (A) *Clostridium perfringens* and (B) *Clostridium sordelli. HS conditions*: Sample: 1.0 mL of the aqueous culture to which 500 mg of NaHSO₄ was added. Equilibration at 120 °C for 20 minutes. *GC conditions*: Column: 15 m × 0.32 mm I.D. fused-silica, open-tubular, coated with FFAP ("free fatty acid") stationary phase; film thickness: 1 μm. Column temperature: programmed from 130 °C to 200 °C at 6 °C/min. Carrier gas: Nitrogen, 243/122 kPa, with the high pressure carrier gas accessory in the Perkin-Elmer HS-100 Automatic Headspace Sampler; split sampling, split 1:10. Flame-ionization detector. Attenuation × 16. *Peaks: 1* = acetic acid, *2* = propionic acid, *3* = 2-methylpropanoic acid, *4* = *n*-butanoic acid, *5* = 3-methylbutanoic acid, *6* = 4-methylpentanoic acid, 7 = not identified.

different culture media be used and the results compared [16]. The evaluation of this combination of different media assures positive identification.

In addition to the volatile fatty acids, other less volatile acids are also present in the culture of bacteria, and their inclusion in the evaluation can further strengthen the identification. To obtain a more complete profile of the fatty acids present, however, these substances now must be converted to their more volatile methyl ester. As described by Heitefuss and co-workers, this esterification can be carried out in situ, in the headspace vial prior to analysis [17]. Figure 8-9 shows typical results from *Bacteroides fragilis.* Heitefuss included three procedures in a general identification procedure, involving the static HS-GC analysis of volatile fermentation products and dicarboxylic acid methyl esters, and the direct GC analysis of long chain fatty acid methyl esters after extraction and esterification [18]. The combined method determines a total of 48 components and can be evaluated by means of a special computer program (BIS = bacteria identification system), which also allows comparison and correlation to 1000 reference strains; however, unambiguous identification still requires additional bacteriological tests. For example, with this computer-aided identification program it was possible to differentiate between the quite similar species *Bacillus cereus* and *Bacillus anthracis,* grown in a synthetic culture medium [19].

Automated static HS-GC is also ideally suited for rapid diagnosis of urinary tract infections and for screening purposes in a hospital to handle a high number of samples or, at least, to eliminate the bulk of uninfected samples. For example, Coloe [20] has demonstrated that certain bacteria can be characterized even quantitatively, by a distribution plot ("scatter plot") of the peak area of ethanol and dimethyl sulfide, measured by HS-GC, versus the number of viable organisms present in unit volume.

HS-GC measurements also offer the possibility of testing for any resistance of microorganisms to certain antibiotics. Since the increasing degree of resistance of bacteria to antibiotics is a growing problem, this type of investigation can help in the search for the most efficient antibiotic for a particular patient. Figure 8-10 illustrates such a test for the impact of amoxycillin using the formation of ethanol E by anaerobes, grown in a universal culture medium, as a key compound. The sample in the series of Figure 8-10B was amoxycillin resistant: thus, the addition of amoxycillin had no effect on ethanol production and therefore did not change the pattern (peak *4* vs. peak *3*). On the other hand, in the series of Figure 8-10A, the addition of amoxycillin killed a significant part of the bacteria present and correspondingly, the size of the ethanol peak was reduced [21].

Finally, an interesting application of HS-GC in the field of microbiology is the determination of phosphine (PH_3) in the headspace of certain bacteria and the distinction of this gas from hydrogen sulfide [22]. Unlike the examples just given, phosphine is not a metabolic product of the nutrient used by the bacteria; rather the bacteria probably detoxify their media (e.g., sludges, soil, harbor sediments), eliminating some phosphor-containing compounds (e.g.,

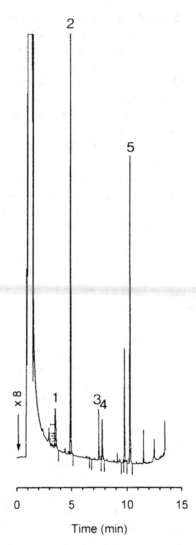

Figure 8-9. Chromatogram of the headspace of the fermentation broth containing *Bacteriodes fragilis*, by HS-GC. *HS conditions*: Sample: 10 μL aqueous solution to which 10 μL of saturated $NaHSO_4$ solution and 10 μL methanol were added. Equilibration at 120 °C for 20 minutes; full evaporation technique. Sample transfer time: 6 seconds. *GC conditions*: Column: 30 m × 0.32 mm I.D. fused-silica, open-tubular, coated with Stabilwax DA poly(ethylene glycol) stationary phase; film thickness: 0.25 μm (Restek). Column temperature: isothermal at 70 °C for 3 minutes, then programmed at 6 °C/min to 120 °C, and from there at 30 °C/min to 200 °C. Carrier gas: Nitrogen, with the high pressure carrier gas accessory in the Perkin-Elmer HS-100 Automatic Headspace Sampler; splitless sampling. Flame-ionization detector. *Peaks*: methyl esters of *1* = pyruvic acid, *2* = lactic acid, *3* = oxalic acid, *4* = malonic acid, *5* = succinic acid.

Source: Reproduced from ref. 17 with permission of S. Heitefuss et al. and *Journal of Chromatography*.

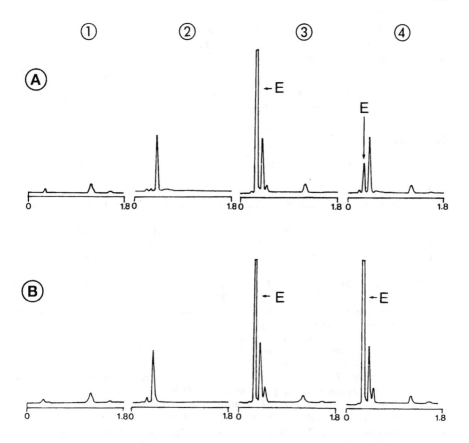

Figure 8-10. Rapid diagnosis of urinary tract infection and the impact of amoxycillin on the formation of ethanol from urine specimens, measured by HS-GC [20]. The cultures were infected by *Escherichia* (or *Klebsiella*, *Citrobacter*, or *Enterobacter*) species. (A) Amoxycillin-susceptible sample. (B) Amoxycillin-resistant sample. *1* = medium blank, *2* = nonincubated urine, *3* = urine culture, *4* = urine culture + amoxycillin. *HS conditions*: Sample: 1.1 mL urine cultures in 1.1 mL arabinose/methionine medium; 3.5-hour incubation at 60 °C. A 2 mL aliquot was transferred to the headspace vial, 3 g K_2CO_3 was added and thermostatted at 60 °C. Medium blank was obtained by processing 1.1 mL uninoculated medium, mixed with 1.1 mL of sterile water. Nonincubated urine results were from the analysis of 1:1 urine specimen and water. *GC conditions*: Column: 2 m × 1/8 in O.D. packed, containing 0.4% Carbowax 1500 poly(ethylene glycol) stationary phase on graphite 60/80 mesh. Column temperature: 115 °C, isothermal. Flame-ionization detector. *E* = ethanol peak.

hypophosphites used originally as fungicides) by reduction to the gaseous phosphine. Without the possibility of direct headspace sampling, such measurements would be fairly complicated.

These examples show the usefulness of headspace–gas chromatography in the qualitative characterization of natural samples, in the use of hyphenated systems, and in clinical microbiology.

References

1. Data Sheet No. HSA-11: *Water Pollution Analysis.* Bodenseewerk Perkin-Elmer & Co., Überlingen, 1977.
2. J. Widomski and W. Thompson, *Chromatogr. Newsl.* **7**, 31–34 (1979).
3. A. J. Taylor, in B. Kolb (editor), *Applied Headspace Gas Chromatography,* Heyden & Sons, London, 1980; pp. 140–154.
4. D. A. M. Mackay, D. A. Lang, and M. Berdick, *Anal. Chem.* **33**, 1369–1374 (1961).
5. R. G. Buttery and R. Teranishi, *Anal. Chem.* **33**, 1440–1441 (1961).
6. H. H. Hill and D. G. McMinn, *Detectors for Capillary Chromatography,* Wiley, New York, 1992; pp. 251–296, 327–353.
7. A. Shinohara, A. Sato, H. Ishii, and N. Onda, *Chromatographia,* **32**, 357–364 (1991).
8. Y. Shirane, *Anal. Views (Japan)* **2**(2), 7–12 (1993).
9. G. L. McClure, in P. B. Coleman (editor), *Practical Sampling Techniques for Infrared Analysis,* CRC Press, Boca Raton, FL, 1993; pp. 165–215.
10. A. Rau and H. Görtz, *Chromatographia,* **28**, 631–638 (1989).
11. G. L. McClure, *Paper No. 756,* 36th Pittsburgh Conference on Analytical Chemistry and Applied Spectroscopy, New Orleans, LA, February 27–March 1, 1985.
12. G. L. McClure and P. R. Roush, *Paper No. 469,* 37th Pittsburgh Conference on Analytical Chemistry and Applied Spectroscopy, Atlantic City, NJ, March 10–14, 1986.
13. Personal information from Dr. G. L. McClure (Perkin-Elmer Corp., Norwalk, CT).
14. P. M. Mitruka, *GC Applications in Microbiology and Medicine,* Wiley, New York, 1973.
15. L. V. Holdeman, E. P. Cato, and W. E. C. Moore, *Anaerobe Laboratory Manual,* 4th ed., Virginia Polytechnic Institute & State University, Blacksburg, VA, 1977.
16. H. S. H. Seifert, H. Böhnel, S. Giercke, A. Heine, D. Hoffmann, V. Sukop, and D. H. Boege, *Int. Lab.,* July/August 1986, pp. 46–56.
17. S. Heitefuss, A. Heine, and H. S. H. Seifert, *J. Chromatogr.* **532**, 374–378 (1990).
18. S. Heitefuss, *Untersuchungen zur Identifizierung von aeroben, anaeroben und fakultativ anaeroben Bakterien mit gas chromatographischen Methoden.* Institut für Pflanzenbau und Tierhygiene in den Tropen und Subtropen, Göttingen Universität, Erich Goetze Verlag, 1991.
19. D. Lawrence, S. Heitefuss, and H. S. H. Seifert, *J. Clin. Microbiol.* **29**, 1508–1512 (1991).
20. P. J. Coloe, *J. Clin. Pathol.* **31**, 365–369 (1978).
21. Courtesy of Dr. N. J. Hayward (Alfred Hospital, Prahran, Victoria, Australia).
22. U. Brunner, T. G. Chasteen, P. Ferloni, and R. Bachofen, *Chromatographia,* **40**, 399–403 (1995).

9

Special Measurements

In addition to the direct analysis of a sample for its quantitative and/or qualitative composition, headspace–gas chromatography can be used for physicochemical measurements. There is in principle no difference as compared to the application of HS-GC for quantitative analysis, except for the type of calibration. The concentrations for quantitative analysis are usually given in the units mass per volume, while molar concentrations are required for thermodynamic measurements.

The physicochemical measurements permit the determination of various characteristic values and functions of gas–liquid phase distribution systems. From pure compounds, vapor pressures and related functions can be derived. Equilibrium HS-GC also facilitates the study of adsorption measurements in gas–solid systems and the determination of the rate of release of a volatile analyte from a solid sample. With respect to the validation of the GC system, the linearity and the detection limit of GC detectors are important practical measurements.

9.1 Determination of Vapor Pressures

When the pure analyte is in the vial, either as a liquid or solid sample, its vapor pressure can be established from the results of the headspace analysis. If n_i is the number of moles of the analyte present, its concentration in the headspace of the vial (expressed as molar concentration) will be n_i/V_G (V_G is the volume

251

of the headspace), and this is proportional to the obtained peak area (cf. eq. 2.17):

$$A_i = f_i \cdot C_G = f_i \cdot (n_i/V_G) \tag{9.1}$$

However, based on the general gas law,

$$p_i = (n_i/V_G) \cdot R \cdot T \tag{9.2}$$

where p_i is the vapor pressure of the pure analyte, R is the gas constant, and T is the absolute temperature. It follows from eqs. 9.1 and 9.2 that

$$A_i = \frac{f_i \cdot p_i}{R \cdot T} \tag{9.3}$$

Equation 9.3 permits the determination of the vapor pressure of the pure compound [1–4] and can also be used for the determination of the common vapor pressure of complex mixtures (e.g., diesel oil or crude oil [5]). Calibration must be carried out by using an external vapor standard. However, vapor pressure data at ambient temperature are mainly required, and it is difficult to prepare such external standards particularly for low vapor pressures (<10 Pa) and far below the boiling point of these compounds, since the necessary vapor standards at the corresponding low concentrations must be prepared by means of the total vaporization technique.

These problems with absolute calibration may be circumvented by using a reference compound with known vapor pressure, under the same conditions as the analyte, whereupon the vapor pressure is calculated from the obtained peak areas and from the molecular weights of the analyte and the reference compound, using the proper detector response factor value [6]. Normal paraffins, of which the vapor pressure at different temperatures is well documented, can conveniently be used as reference compounds of this type, and they comprise a wide range of vapor pressures. Any other well-documented compound can of course be used equally well.

To understand this method of calculation, we first consider the relationship of molecular weight to volume for the vapors of a compound. Assuming that only the vapor of the compound of interest is present, the vapor of an amount corresponding to the gram molecular weight (M_i) of the compound would occupy the molar volume V_{mole} [22.414 L at 0 °C and 101,325 Pa (1 atm)]. Thus, the number of moles present in a volume of V_G will be $n_i = (V_G/V_{mole}) \cdot M_i$. However, we also have other gases (e.g., air) present in the vial. Therefore, the actual number of moles present will depend on the partial pressure of the compound in the vial (p_i) versus the total pressure (p_{total}) in the vial:

$$n_i = \frac{p_i}{p_{total}} \cdot \frac{V_G}{V_{mole}} \cdot M_i \tag{9.4}$$

If in a separate vial we have the reference compound present under identical conditions, we can express n_{ref}, the number of moles of this compound present in the headspace volume of the vial, in the same way as eq. 9.4, by substituting its partial pressure p_{ref} for p_i and its molecular weight M_{ref} for M_i:

$$n_{ref} = \frac{p_{ref}}{p_{total}} \cdot \frac{V_G}{V_{mole}} \cdot M_{ref} \tag{9.5}$$

Dividing eq. 9.4 by eq. 9.5 gives:

$$\frac{n_i}{n_{ref}} = \frac{p_i}{p_{ref}} \cdot \frac{M_i}{M_{ref}} \tag{9.6}$$

Going now back to eq. 9.1, we can write it for the analyte and the reference compound:

$$A_i = f_i \cdot (n_i / V_G) \tag{9.7}$$

$$A_{ref} = f_{ref} \cdot (n_{ref} / V_G) \tag{9.8}$$

Thus:

$$\frac{A_i}{A_{ref}} = \frac{f_i}{f_{ref}} \cdot \frac{n_i}{n_{ref}} \tag{9.9}$$

Combining eqs. 9.6 and 9.9 gives:

$$\frac{A_i}{A_{ref}} = \frac{f_i}{f_{ref}} \cdot \frac{p_i}{p_{ref}} \cdot \frac{M_i}{M_{ref}} \tag{9.10}$$

Here f_i/f_{ref} can be replaced by RF, the detector response of the analyte relative to the reference compound; this factor can be determined in the usual way, by normal syringe injection of both compounds, and does not need headspace sampling. As a conclusion, the vapor pressure of the analyte can be established from the known and measured values.

$$p_i = p_{ref} \cdot RF \frac{A_i}{A_{ref}} \cdot \frac{M_{ref}}{M_i} \tag{9.11}$$

This procedure is applied in Example 9.1 to determine the vapor pressures of 2,4,6-trichloro-1,3,5-triazine (TCTA) in the temperature range of 20 to 150 °C utilizing n-tridecane (n-C_{13}) as the reference compound. The results are shown in Figure 9-1, where the logarithms of the vapor pressures are plotted versus $1/T$. The compound TCTA is used as a precursor for herbicide synthesis.

Example 9.1

Determination of the detector response factor RF. Approximately equal amounts of TCTA (10.1 mg) and n-C_{13} (10.0 mg) were dissolved in 1 mL acetone, and 1 μL of this solution was injected five times into the gas chromatograph. The analysis was carried out with a packed column (6 ft $\times$ 1/8 in. O.D. stainless steel column, 4% dimethyl silicone on Chromosorb G, 60/80 mesh) at 100 °C and a thermal-conductivity detector. A detector response factor of $RF = 1.682$ ($\pm 0.8\%$ RSD, $N = 5$) for TCTA was determined from the resulting area values in the usual way.

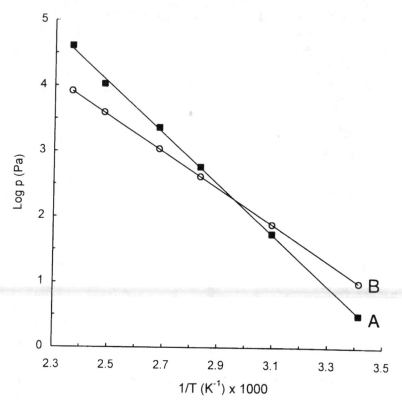

Figure 9-1. Temperature function of the vapor pressures p of 2,4,6-trichloro-1,3,5-triazine (A) and of n-tridecane (B) as the reference compound. The pertinent data are listed in Table 9-1.

Headspace procedure. Each compound (TCTA and n-C_{13}) was dispensed in several headspace vials and analyzed under the chromatographic conditions just given, but at various sample temperatures. The pertinent data are listed in Table 9-1 [7] Sample size was 50 mg at each compound.

9.2 Determination of Activity Coefficients

In headspace–gas chromatography we are investigating binary equilibrium systems in which the basic relationship is described by eq. 2.26:

$$p_i = p_i^o \cdot \gamma_i \cdot x_{S(i)} \tag{2.26}$$

Here, $x_{S(i)}$ is the mole fraction of the analyte in a solution, p_i is its partial pressure over the solution, p_i^o is the vapor pressure of the pure analyte (when $x_{S(i)} = 1$) and γ_i is the so-called activity coefficient of the analyte; the pressures

Table 9-1   Determination of the vapor pressure of 2,4,6-trichloro-1,3,5-triazine
(TCTA) using n-tridecane (n-C$_{13}$) as reference system according to
eq. 9.11: $RF = 1.682$; molecular weights: TCTA: $M_i = 184.43$;
n-C$_{13}$: $M_{ref} = 213.36$

		Peak area (counts)		Vapor pressure (Pa)	
		Reference n-C$_{13}$	Analyte TCTA	Reference* n-C$_{13}$	Analyte TCTA
°C	$(1/T) \, 10^3$	A_{ref}	A_i	p_{ref}	p_i
20	3.4111	0.1279	0.0207	10.019	3.155
50	3.0960	0.8202	0.3023	76.353	54.760
80	2.8329	4.5710	3.2752	411.975	574.407
100	2.6810	11.4020	12.4981	1090.232	2325.431
130	2.4814	38.4408	53.7720	3916.39	10660.35
150	2.3641	67.7535	169.421	8303.51	40403.53

Linear regression analysis of $\log p = \dfrac{q}{T} + b$:

		Reference	Analyte
Slope	a:	$-2{,}786.52$	$-3{,}871.67$
Intercept	b:	10.5078	13.7165
Correlation coefficient	r:	-0.99999	-0.99958

* From ref. 7 and recalculated to pascals with 1 torr = 133 Pa.

p_i and p_i^0 are temperature-dependent values. In a dilute solution (like those used in general for quantitative measurements) one can usually assume that the activity coefficient is constant, thus the partial pressure — and the corresponding peak area — are directly proportional to concentration. However, this is not the case for concentrated mixtures, where the activity coefficient is a concentration-dependent function. The activity coefficient is an important value, and knowledge of it is necessary in a number of physicochemical calculations. Headspace–gas chromatography presents an elegant method for its determination.

It is evident from the discussion in Section 9.1 that in a given system, the peak area obtained in the headspace measurement of a mixture A_i is proportional to the partial vapor pressure of the analyte p_i in the vial over the mixture [8]:

$$A_i = c \cdot p_i \tag{9.12}$$

where c is a calibration factor. Combining this relationship with the basic relationship given in eq. 2.26 gives [9]:

$$A_i = c \cdot p_i^0 \cdot \gamma_i \cdot x_{S(i)} \tag{9.13}$$

The area A_i was obtained from analyzing the headspace of a (binary) mixture, and it is proportional to the partial vapor pressure of the analyte over the mixture at equilibrium. Similarly, if we have the pure analyte present in a vial, the peak area obtained A_i^o will be proportional to the vapor pressure of the analyte at the prevailing temperature:

$$A_i^o = c \cdot p_i^o \qquad (9.14)$$

Thus, eq. 9.13 can be written as follows:

$$A_i = A_i^o \cdot \gamma_i \cdot x_{S(i)} \qquad (9.15)$$

From eq. 9.15 the activity coefficient can be expressed:

$$\gamma_i = \frac{A_i}{A_i^o \cdot x_{S(i)}} \qquad (9.16)$$

Thus if we know the molar concentration of a binary mixture, we can establish the activity coefficient of both components. Its value is concentration dependent; as an example, Figure 9-2 plots the activity coefficients of acetone and chloroform at 36.5 °C against the molar concentration of their mixture [9], together with literature data measured by other methods [10]. As seen, the agreement is very good.

Since the peak areas and the vapor pressures are directly related, vapor phase diagrams can be established by plotting the peak areas* against the molar concentration of binary mixtures, which show the deviation from linearity due to the concentration-dependent activity coefficient. Figure 9-3 shows such a plot for the chloroform/acetone mixture [9]. This method is not restricted to binary mixtures but can be extended to multicomponent mixtures: plots for the ternary mixture acetone/chloroform/methanol are shown in Figure 9-4 [11].

9.3 Determination of Related Physicochemical Functions

The activity coefficient has a key function for the thermodynamics of mixed phases. Since

$$\gamma_i \cdot x_i = A_i / A_i^o \qquad (9.17)$$

the peak area ratio can be used in calculations involving the activity coefficient. For example, the *partial free molar energy of mixing* ΔG_i^M can be expressed as follows:

$$\Delta G_i^M = R \cdot T \cdot \ln(A_i / A_i^o) \qquad (9.18)$$

---

* For such plots, the peak areas of one component must be normalized by its response factor to compensate for the different response for the individual components of the mixture.

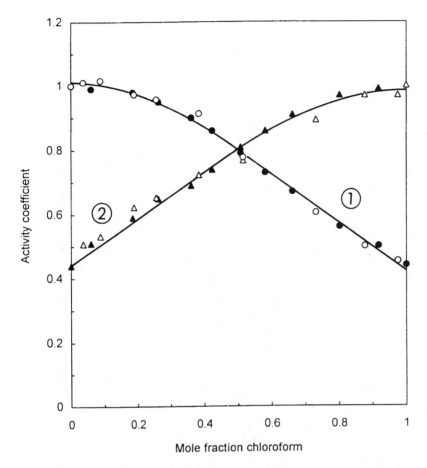

Figure 9-2. Activity coefficients of acetone *(1)* and chloroform *(2)*, as a function of the molar concentrations in the binary mixture, at 36.5 °C [9]. Open symbols: results of HS-GC measurement; solid symbols: data from ref. 10.

while ΔG^M, the *total free energy of mixing* a binary system from its pure liquid components, can be calculated from:

$$\Delta G^M = R \cdot T \cdot [x_1 \cdot \ln(A_1/A_1^0) + x_2 \cdot \ln(A_2/A_2^0)] \tag{9.19}$$

In a similar way the corresponding excess functions ΔG^E can be derived from the activity coefficients [9].

Such physicochemical values are important for chemical engineering applications, and a high number of such data are required for technical purposes. Automated HS-GC offers good capabilities to process the many systems and to collect the necessary high number of data.

Hachenberg and Schmidt [12] have applied HS-GC to the selection of

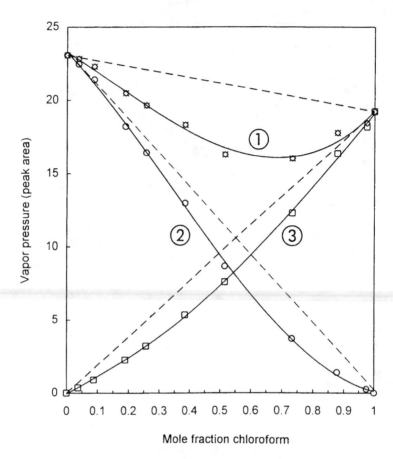

Figure 9-3. Partial and total pressure relationship of chloroform/acetone mixtures, at 36.5 °C [9]. Pressures are expressed as arbitrary units of the corresponding peak areas from HS-GC measurements. Broken lines: ideal relationship (for $\gamma = 1.000$). 1 = total pressure; 2 = partial pressure of acetone 3 = partial pressure of chloroform.

suitable solvents for extractive distillation by determining the improvement factor IF for the separation of two components 1 and 2 if a third compound 3 is added:

$$IF = \frac{(A_1/A_2)_{123}}{(A_1/A_2)_{12}} \tag{9.20}$$

where

$(A_1/A_2)_{123}$ = area ratio of components 1 and 2 with component 3 present

$(A_1/A_2)_{12}$ = area ratio of components 1 and 2 without component 3

This improvement effect is apparently caused by intermolecular interaction and can be expressed by the underlying activity coefficients. If the activity coeffi-

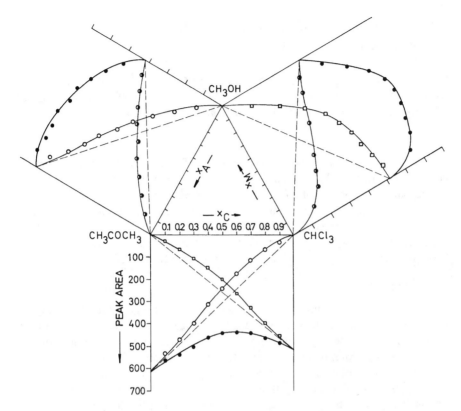

Figure 9-4. Partial and total pressure relationship of the ternary system methanol/ acetone/chloroform, at 50 °C. Concentrations in the condensed (liquid) phase are given as mole fraction (x), while in the gas phase as corrected values of the peak areas. x_A = mole fraction of acetone, x_C = mole fraction of chloroform, x_M = mole fraction of methanol.

Source: Reproduced from ref. 11 with permission of *Ber. Bunsenges. Phys. Chem.*

cients of compounds 1 and 2 are determined, the influence of the added compound 3 on the relative volatility of both compounds 1 and 2 can be expressed by the selectivity S [12]:

$$S = (\gamma_1/\gamma_2)_{123} \tag{9.21}$$

In a similar way the intermolecular interactions in various systems were investigated [13] to calculate the group interaction parameters for the group contribution method UNIFAC. Such vapor–liquid equilibria investigations were applied also for polymer solutions in terms of modified UNIFAC-FV models [14,15].

A number of other physicochemical parameters can be established with the help of HS-GC; interested readers are referred to the specific chapters of the textbooks [12, 16,17] and the references cited there.

9.4 Determination of Phase Distribution (Partition Coefficient)

The phase distribution of a volatile compound between the gas phase and the condensed phase is usually described by the partition coefficient K, which is analogous to the partition coefficient in a gas chromatographic column:

$$K = C_S/C_G \tag{2.11}$$

where C_S and C_G are the concentrations of the analyte in the sample (condensed) phase and the gas phase (headspace), respectively.

The partition coefficient is identical to the "solubility coefficient" for solid polymers as the condensed phase [18]. In the case of phase distribution between air and water, the Henry's law constant expressed in [atm · m³/mol] is often preferred; however, the dimensionless Henry's law constant H can also be defined as the reciprocal of the partition coefficient [19,20]:

$$H = C_G/C_S = 1/K \tag{9.22}$$

The partition coefficient is a fundamental value in gas chromatography in general, and also in headspace sampling, where it is one of the two factors (the other being the phase ratio) influencing the sensitivity of the determination. In many cases, therefore, it may be important to know the value of K.

There are a number of ways in which the partition coefficient of an analyte between a stationary (liquid) phase and the carrier gas can be determined, and such measurements have been described since the inception of gas chromatography. HS-GC is a convenient method for such measurements. Probably the earliest work utilizing HS-GC for this purpose was that of Rohrschneider [21]. More recently, two methods were developed for the determination of gas–liquid partition coefficients utilizing HS-GC: the vapor phase calibration (VPC) method [22] and the phase ratio variation (PRV) method [23], summarized in Sections 9.4.1 and 9.4.2, respectively. A third possibility in which multiple headspace extraction measurements are utilized is also discussed.

The actual examples to be given refer to gas–water systems. We emphasize, however, that the techniques can be used with any solvent and also with solid samples. They can also serve for the determination of the partition coefficient between a stationary (liquid) phase used in gas chromatography and the carrier gas, and the referenced publications [22,23] give examples for such determinations.

9.4.1 The Vapor Phase Calibration (VPC) Method

If small amounts of a pure analyte are introduced into a headspace vial that is thermostatted at an elevated temperature, the full amount of the introduced analyte will evaporate. This is the total vaporization technique (covered in

Section 4.6.1), and it results in one-phase system: the vial containing the vapor of the analyte plus air. Such a sample can therefore be used as a calibration standard. Another vial contains the solvent to be used for the determination, and the same amount of the analyte is added to it (conveniently by syringe injection into the vial containing the solvent). The same method enabled the determination of Ostwald's solubility coefficients of dissolved gases in transformer oil [23]. The principles of the technique are summarized next.

To the vial having a volume of V_V we added V_S volume of the solvent; thus the volume of the headspace gas is $V_G = V_V - V_S$. We add the amount W_o of the analyte into the vial; after equilibration, the respective amounts of the analyte in the two phases are W_S and W_G, and the two concentrations are C_S and C_G.

When analyzing an aliquot of the headspace, peak area A_G is obtained, which is proportional to the headspace concentration; the proportionality constant is f:

$$C_G = f \cdot A_G \tag{9.23}$$

Since $C_G = W_G/V_G$, we can write

$$W_G = f \cdot A_G \cdot V_G \tag{9.24}$$

Into a separate vial containing no solvent (the "calibration vial"), the same amount of analyte that went into the sample vial (W_o) was introduced and evaporated completely. Its concentration in the vial will be C_C:

$$C_C = W_o/V_V \tag{9.25}$$

Analyzing an aliquot of this gas under conditions identical to those of the sample vial, peak area A_C is obtained, which will have the same proportionality constant f:

$$C_C = f \cdot A_C \tag{9.26}$$

hence

$$W_o = f \cdot A_C \cdot V_V \tag{9.27}$$

The gas–liquid partition coefficient K of the analyte is:

$$K = \frac{C_S}{C_G} = \frac{W_S}{W_G} \cdot \frac{V_G}{V_S} \tag{2.12}$$

We know W_G but we also need W_S; this can be established from the material balance of

$$W_o = W_S + W_G \tag{2.10}$$

substituting eq. 9.27 for W_o and eq. 9.24 for W_G:

$$W_S = W_o - W_G = f \cdot A_C \cdot V_V - f \cdot A_G \cdot V_G = f \cdot (A_C \cdot V_V - A_G \cdot V_G) \tag{9.28}$$

Substituting this and eq. 9.24 into eq. 2.12 we obtain:

$$K = \frac{f \cdot (A_C \cdot V_V - A_G \cdot V_G) \cdot V_G}{f \cdot A_G \cdot V_G \cdot V_S} = \frac{A_C \cdot V_V - A_G \cdot V_G}{A_G \cdot V_S} \tag{9.29}$$

Although in this derivation the amount of analyte is used, in practice we usually add it by *volume*. However, since we have W_S/W_G in eq. 2.12, the density needed to convert volume to mass will cancel out.

In Example 9.2, we determine the partition coefficient of methyl ethyl ketone (MEK) in a gas–water system at 70 °C.

Example 9.2

To a vial ($V_V = 22.3$ mL) containing $V_S = 3.0$ mL of water, 2.0 μL of MEK is added by means of a microsyringe; $V_G = 22.3 - 3.0 = 19.3$ mL. An identical volume of MEK is added into an empty vial and the vials are thermostatted at 70 °C. The following peak area values were measured: $A_G = 4868$; $A_C = 35,202$.

$$K = \frac{35,202 \times 22.3 - 4868 \times 19.3}{4868 \times 3.0} = 47.32$$

If the partition coefficient is high, then A_G will be much smaller than A_C, which may even bring the A_C values out of the linear range. In this case the VPC method may be modified in that the amount (volume) of pure analyte added into the (empty) calibration vials is *different* from the amount (volume) of the analyte added into the sample vial containing the solvent.

If we add W_C instead of W_o to the calibration vial, the peak area obtained will be A'_C and the concentration in the vial will be

$$C'_C = \frac{W_C}{V_V} = f \cdot A'_C \tag{9.30}$$

$$W_C = f \cdot A'_C \cdot V_V \tag{9.31}$$

The respective peak area that would correspond to W_o can be calculated based on the proportionality of peak area and concentration:

$$\frac{W_o}{W_C} = \frac{f \cdot A_C}{f \cdot A'_C} \tag{9.32}$$

$$A_C = A'_C \cdot \frac{W_o}{W_C} = A'_C \cdot r \tag{9.33}$$

where:

$$r = W_o/W_C \tag{9.34}$$

Thus:

$$K = \frac{V_V \cdot A_C \cdot r - V_G \cdot A_G}{A_G \cdot V_S} \tag{9.35}$$

This modified VPC method is illustrated in Example 9.3 as we determine the partition coefficient of MEK, but now at 45 °C.

Example 9.3

The pertinent data are as follows: $V_V = 22.3\,\text{mL}$; $V_S = 4.0\,\text{mL}$; $V_G = 22.3-5.0 = 17.3\,\text{mL}$. We added $10.0\,\mu\text{L}$ MEK to the sample vial and $2.0\,\mu\text{L}$ MEK into the calibration vial; thus $r = 10/2 = 5$. The obtained peak areas are : $A_G = 2,235$ and $A_C' = 14,540$

$$K = \frac{(22.3 \times 14,540 \times 5) - (17.3 \times 2235)}{2235 \times 5} = 141.6$$

The partition coefficient of MEK at $45\,°\text{C}$, determined by the PRV method (see Section 9.4.2) is given in the literature [24] as 144.7, a difference of 2%.

The accuracy of determinations by the VPC method can be improved by carrying out a number of measurements with different analyte amounts and either taking the mean of the results or utilizing linear regression analysis of the peak area versus analyte amount values. For the sample series, we can write

$$A_G = a_G \cdot W_G + b_G \tag{9.36}$$

where a_G and b_G are regression constants. Since W_G is proportional to W_o, we can use its value in eq. 9.37:

$$A_G = a_G \cdot W_o + b_G \tag{9.37}$$

We can write a similar relationship for the calibration series:

$$A_C = a_C \cdot W_o + b_C \tag{9.38}$$

where a_C and b_C are again regression constants.

Since the intercept values will be very small relative to the actual peak area, we can substitute $a_C \cdot W_o$ and $a_G \cdot W_o$ into eq. 9.39 for A_C and A_G in eq. 9.29, respectively.

$$K = \frac{a_C \cdot W_o \cdot V_V - a_G \cdot W_o \cdot V_G}{a_G \cdot W_o \cdot V_S} = \frac{a_C \cdot V_V - a_G \cdot V_G}{a_G \cdot V_S} \tag{9.39}$$

This linear regression calculation is illustrated next, using the same sample as in Example 9.2 (MEK at $70\,°\text{C}$) but now having different analyte volumes.

Example 9.4

The following four data sets were obtained when adding 2.0, 3.0, 4.0, and $5.0\,\mu\text{L}$ of the MEK to the vials; other values are the same as in Example 9.2. The peak area values are:

$W_o\,\mu\text{L}$	2.0	3.0	4.0	5.0
A_G	4,868	7,005	9,207	11,434
A_C	35,202	51,357	66,903	83,659

The results of the linear regression analysis are:

$$a_G = 2,190 \qquad b_G = 463.5 \qquad r_G = 0.99996$$
$$a_C = 16,091.1 \qquad b_C = 2,960.0 \qquad r_C = 0.99990$$

This value differs by 1.8% from the one calculated in Example No. 9.1.

Earlier we mentioned that in the case of high partition coefficients, the A_C values may be out of the linear range. This can be checked easily by the linear regression analysis of the W_o-versus-A_C values.

9.4.2 The Phase Ratio Variation (PRV) Method

The PRV method [24] is based on the relationship between the reciprocal peak area and the phase ratio of the vial containing the sample solution. Linear regression analysis is applied to the results obtained from a number of vials containing the same sample solution but with a wide range of phase ratios β.

9.4.2.1 Principles

According to the fundamental relationship of headspace–gas chromatography defined in Chapter 2 (eq. 2.19):

$$C_G = \frac{C_o}{K + \beta} \tag{2.19}$$

where C_o is the analyte concentration in the original sample and C_G is its concentration in the headspace at equilibrium conditions; K is the partition coefficient, and β is the phase ratio of the vial. Taking the reciprocals of both sides of this equation we obtain:

$$\frac{1}{C_G} = \frac{K + \beta}{C_o} = \frac{K}{C_o} + \frac{\beta}{C_o} \tag{9.40}$$

If we have the same sample solution (with the same analyte concentration) but in different volumes in a number of headspace vials that are equilibrated at the same temperature K and C_o will be the same for each vial; the only variables are C_G and β. In other words, eq. 9.40 corresponds to a linear equation of the $y = ax + b$ type, where

$$y = 1/C_G$$

$$x = \beta$$

If we write eq. 9.40 in the following form:

$$\frac{1}{C_G} = a \cdot \beta + b \tag{9.41}$$

its slope will be $a = 1/C_o$ and its intercept $b = K/C_o$; the partition coefficient can be calculated from these two values:

$$K = b/a \tag{9.42}$$

Equation 9.40 expresses the headspace concentration as a function of the other values. However, what we are measuring is the peak area obtained when analyzing an aliquot of the headspace, which is proportional to the equilibrium headspace concentration:

$$A_G = f \cdot C_G \tag{9.43}$$

$$C_G = A_G/f \tag{9.44}$$

Substituting this expression into eq. 9.40 we obtain:

$$\frac{1}{A_G} = \frac{K + \beta}{f \cdot C_o} = \frac{K}{f \cdot C_o} + \frac{1}{f \cdot C_o} \cdot \beta \tag{9.45}$$

This relationship again corresponds to a $y = a'x + b'$ type of linear regression equation, where:

$$y = 1/A_G$$

$$x = \beta$$

and the slope and intercept of eq. 9.45 can be expressed as follows:

$$a' = \frac{1}{f \cdot C_o}$$

$$b' = \frac{K}{f \cdot C_o}$$

The partition coefficient can again be calculated from these two values:

$$K = b'/a' \tag{9.46}$$

In other words, by linear regression analysis of β against $1/A_G$, the slope (a') and intercept (b') of the linear plot corresponding to eq. 9.45 can be established. From these values, the partition coefficient can be calculated according to eq. 9.46.

In the next example we utilize the PRV method for the determination of the partition coefficient of toluene in an air/water system, at 60 °C.

Example 9.5

An aqueous solution of toluene, with a concentration of 10 μL/mL, was prepared, and different volumes of this solution were placed into four headspace vials. The vials were thermostatted at 60 °C and then analyzed by HS-GC. The following results were obtained:

Vial no.	Phase ratio β	Peak area values obtained A_G
1	21.300	18,140
2	10.150	35,115
3	6.433	50,550
4	4.575	65,305

Linear regression analysis of β versus $1/A_G$ resulted in the following values:

$a' = 0.0023797$

$b' = 0.0044162$

$r = 0.99999$

The calculated value of the partition coefficient is (eq. 9.46):

$$K = \frac{0.0044162}{0.0023797} = 1.86$$

9.4.2.2 Limitation of the PRV Method

This method for the determination of the partition coefficient depends on peak area differences, resulting from changing the phase ratio values. Obviously, the accuracy of the calculations can be improved by having differences as large as possible in the peak area values. However, in the case of large K values, the relative area differences become small. This is clear if we write eq. 9.45 for two vials and then take the ratio of the two equations:

$$\frac{1}{A_{G1}} = \frac{K + \beta_1}{f \cdot C_o} \qquad (9.45a)$$

$$\frac{1}{A_{G2}} = \frac{K + \beta_2}{f \cdot C_o} \qquad (9.45b)$$

$$\frac{A_2}{A_1} = \frac{K + \beta_1}{K + \beta_2} \qquad (9.47)$$

Obviously if K is small relative to the phase ratios, the value of A_2/A_1 will approach that of β_1/β_2. On the other hand, if K is much larger than the phase ratio values, then A_2/A_1 will approach unity.

if $K \ll \beta$,   then $A_2/A_1 \to \beta_1\beta_2$

if $K \gg \beta$,   then $A_2/A_1 \to 1$

For low K values a good example is toluene, where β_1/β_2 (the ratio of the highest and smallest phase ratios) was $21.30/4.575 = 4.66$, and a fairly wide range of peak areas (65,305/18,140 = 3.60) was obtained (see Example 9.5). On the other hand, for methyl ethyl ketone, whose partition coefficient at 45 °C was calculated as 144.7 [24], the same phase ratio range would give a peak area ratio of only 1.12, and the problem can already be observed from the correlation coefficient for this particular case: it was just 0.99108 [24]. We consider about 1.12, to be as the reasonable minimum for A_2/A_1 which would correspond to a partition coefficient of about 75–100.* This does not mean that the PRV method could not be used outside this limit: however, it must be understood that for higher partition coefficients (where $A_2/A_1 \leqslant 1.12$) the PRV method will give poorer results.

9.4.3 MHE Methods for the Determination of the Partition Coefficient

McAuliffe, who originally described the basis of the MHE method, has shown [25] how the Henry's law constant determines the extraction effect of a stepwise extraction procedure. If now the MHE technique is carried out by the balanced pressure sampling technique, the partition coefficient (K) can be determined from the area ratio Q characteristic for an MHE measurement series:

$$Q = \frac{A_2}{A_1} = \frac{A_3}{A_2} = \frac{A_{(i+1)}}{A_i} \tag{2.61}$$

where A_1, A_2, A_3, etc. are the peak areas obtained in the consecutive measurements, and i is the number of extraction steps performed. In Section 5.5.4 we expressed the relationship between quotient Q, the partition coefficient K, and the phase ratio of the vial β as follows:

$$Q_s = \frac{K/\beta + \rho}{K/\beta + 1} \tag{5.43}$$

where

$$\rho = p_o/p_h \tag{5.40}$$

In other words, ρ is in the MHE measurement the ratio of the (absolute) pressure in the vial before (p_h) and after (p_o) venting.

The MHE method and the area ratio Q can be utilized in two ways for the determination of the partition coefficient: the first resembles the vapor phase calibration (VPC) method, while the second resembles the phase ratio variation (PRV) method.

---

* For example, for methyl ethyl ketone at 60 °C, $K = 69.2$, and the value of A_2/A_1 for the given phase ratio range is 6795/5545 = 1.23. The correlation coefficient of the linear regression analysis was $r = 0.99834$, better than for the measurement at 45 °C [24].

9.4.3.1 VPC/MHE Method

In Section 5.5.4.2 we saw that for total vaporization (which is the case with the standard in the VPC method), the pressure ratio ρ is equal to the area ratio values (Q_{st}):

$$\rho = Q_{st} \tag{5.44}$$

Substituting eq. 5.44 into eq. 5.43 we obtain:

$$Q_s = \frac{K/\beta + Q_{st}}{K/\beta + 1} \tag{9.48}$$

and from this relationship we can express K as:

$$K = \frac{Q_{st} - Q_s}{Q_s - 1} \cdot \beta \tag{9.49}$$

This method is illustrated here by two examples. The first refers to the determination of the partition coefficient of hydrogen sulfide in an air–crude oil system, at 80 °C; we shall also use this example to calculate the H$_2$S concentration in the crude oil.

Example 9.6

Into vial 1 we added 0.34 mL ($\beta = 64.59$) of the crude oil sample containing H$_2$S, while into vial 2 we injected 50 μL of pure H$_2$S gas; the vial volume was 22.3 mL. Both vials were thermostatted at 80 °C for 60 minutes.

At 20 °C and 101.3 kPa (1 atm) pressure, the molar volume of a gas is:

$$22.414 \cdot \frac{293.16}{273.16} = 24.055 \text{ L}$$

The molecular mass of H$_2$S is 34.08 g; thus, 50 μL of H$_2$S gas at the given conditions corresponds to 70.84 μg H$_2$S.

Table 9-2 lists the results of the MHE measurement, together with the regression data; the corresponding plots are given in Figure 9-5. The following GC conditions were used for the analysis: 25 m × 0.32 mm I.D. fused-silica, porous-layer, open-tubular column coated with Porapak Q type porous polymer. Column temperature: 70 °C; split sampling. Flame-photometric detector (FPD).

Calculation of the partition coefficient (eq. 9.49):

$$K = \frac{0.469683 - 0.601718}{0.601718 - 1} \cdot 64.6 = 21.4$$

From the available data we can also calculate the concentration of H$_2$S in the crude oil. For this we calculate the sum of the peak areas according to eq. 5.28. The total peak area for the external vapor standard is further corrected by the volume correction factor (eq. 5.37).

Example 9.7

The volume correction factor (f_V) is:

$$f_V = \frac{V_V}{V_V - V_S} = \frac{22.3}{22.3 - 0.34} = 1.0155$$

Table 9-2 Determination of the partition coefficient of hydrogen sulfide in an air–crude oil system, at 80 °C, using the VPC method with MHE measurement* (Example 9.6)

	Peak area (counts)	
i	Sample, A_i	External vapor standard, A_{ex}
1	80,411	99,431
2	49,504	46,254
3	29,114	21,888
Linear regression		
Correlation coefficient r	−0.999662	−0.999979
Slope q	−0.50797	−0.75676
$Q = e^{-q}$	0.60172	0.46918
Intercept A^*	81,026	99,148

* For the linear regression plots, see Figure 9-5.

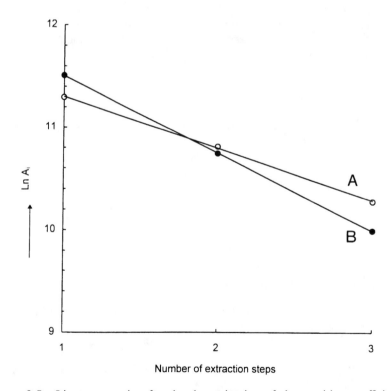

Figure 9-5. Linear regression for the determination of the partition coefficients of hydrogen sulfide in an air–crude oil system at 80 °C, using the VPC method with MHE (Example 9.6). A = oil sample, B = calibration standard.

The sums of the peak areas are:

sample: $\Sigma A_i = \dfrac{81.026}{1.0 - 0.60172} = 203{,}439$

external vapor standard: $\Sigma A_{ex} = \dfrac{99{,}148}{1.0 - 0.46918} = 186{,}784$

The corrected sum of the peak areas of the external standard is (cf. 5.38):

$$\Sigma A_{ex}^x = 186{,}784 \cdot 1.0155 = 189{,}679$$

The amount of H_2S present in the sample is:

$$\dfrac{203{,}439}{189{,}679} \cdot 70.84 = 75.98 \ \mu g$$

representing a concentration of $75.98/0.34 = 223.5 \ \mu g/mL$ crude oil.

The second example deals with the determination of the partition coefficient of styrene monomer in polystyrene, freeze-ground for the measurements. The actual measurements carried out at $120\,°C$ were given in Example 6.1 and in Tables 6-1 and 6-2: here only the final results are repeated. No volume correction factor was used, since with the small volume used (0.2 mL), it would be less than 1%. The value obtained for K should be considered to be only approximate, since the phase ratio was calculated using the bulk volume of the polystyrene sample, without regard to the porosity of the sample that, however, would contribute to the gas phase volume in the vial.

Example 9.8

The following values were obtained for the area ratios Q for styrene monomer:

standard (Table 6-1): $Q_{st} = 0.39727$

sample (Table 6-2): $Q_s = 0.74702$

The volume of the sample was 0.2 mL. Thus the phase ratio is:

$$\beta = \dfrac{22.3 - 0.2}{0.2} = 110.5$$

The partition coefficient of styrene monomer in the polystyrene pellets is (eq. 9.49):

$$K = \dfrac{0.39727 - 0.74702}{0.74702 - 1} \cdot 110.5 = 152.8$$

9.4.3.2 PRV/MHE Method

In this case we analyze the same sample in two different volumes but otherwise, under identical conditions, resulting in the area ratios Q_1 and Q_2. We can write eq. 5.43 for the two cases:

$$Q_1 = \dfrac{K/\beta_1 + \rho}{K/\beta_1 + 1} \tag{9.50a}$$

$$Q_2 = \dfrac{K/\beta_2 + \rho}{K/\beta_2 + 1} \tag{9.50b}$$

From these two equations we can express K:

$$K = \frac{(Q_2 - Q_1) \cdot \beta_1 \cdot \beta_2}{(Q_1 - 1) \cdot \beta_2 - (Q_2 - 1) \cdot \beta_1} \tag{9.51}$$

This method is illustrated here by the determination of the partition coefficient of methyl ethyl ketone (MEK) in a water–air system, at $70\,°C$. Two samples were prepared by adding $3\,\mu L$ of MEK into the vials containing 1.0 and 5.0 mL of water.

Example 9.9

Vials 1 and 2 contained 1.0 and 5.0 mL of water, respectively, representing the phase ratios of $\beta_1 = 21.3$ and $\beta_2 = 3.46$. To each vial $3\,\mu L$ of MEK ($= 2.415\,mg$) was added, and the vials were thermostatted at $70\,°C$ for 60 minutes. The GC analysis was carried out on a $2\,m \times 1/8\,in.$ O.D. packed column containing 15% Carbowax 1500 on Chromosorb W 60/80 mesh, at $60\,°C$, using a flame-ionization detector. Table 9-3 lists the results of the MHE measurement, together with the regression data; the corresponding plots are shown in Figure 9-6. Substituting the respective values of Q and β into eq. 9.51 gives $K = 44.5$.

Comparison of this result with that of Example 9.4 gives a difference of 7.6%. This discrepancy undoubtedly exists because the value of $A_i/A_{(i+1)}$ for the 5.0 mL sample is 1.04, which is much less than the recommended lower limit. This can also be seen from the value of the slope of the regression plot ($q = -0.04$), which is approaching zero.

Table 9-3 Determination of the partition coefficient of methyl ethyl ketone in an air–water system, at $70\,°C$, using the PRV method with MHE measurements.*
(Example 9.9)

	Peak area (counts) for two sample volumes	
i	1.0 mL ($\beta = 21.3$)	5.0 mL ($\beta = 3.46$)
1	5,346	1,470
2	4,341	1,407
3	3,548	1,348
4	2,893	1,296
Linear regression		
Correlation coefficient r	−0.99998	−0.99970
Slope q	−0.20439	−0.04208
$Q = e^{-q}$	0.8151	0.9588
Intercept (A_1^*)	5,338	1,469

* For the linear regression plots, see Figure 9-6.

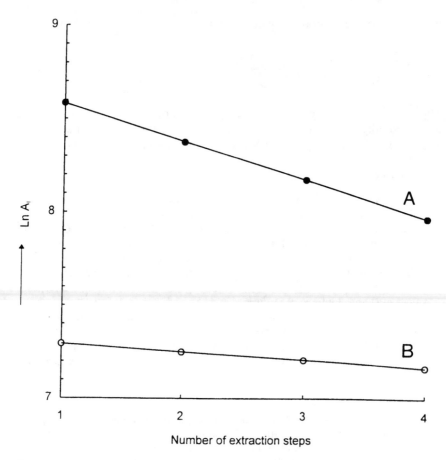

Figure 9-6. Linear regression plots for the determination of the partition coefficients of methyl ethyl ketone in an air–water system at 70 °C, using the PRV method with MHE (Example 9.9). $A = 1.0$ mL sample, $B = 5.0$-mL sample.

9.5 Reaction Constant Measurements

If a volatile compound is involved in a chemical reaction, it is possible to carry out this reaction in a headspace vial and to monitor the volatile compound for kinetic investigations, preferably by the progressive mode (see Section 3.4.2).

The use of HS-GC for such investigations has been discussed in the literature (see, e.g., refs 26–28). An alternative approach to kinetic measurements is the determination of the equilibrium constant of a chemical reaction involving equimolar quantities of components X and Y, forming adduct XY:

$$X + Y \rightleftarrows XY$$

where compound X is a volatile and compound Y a nonvolatile reactant. The equilibrium constant (EC_{XY}) of this reaction is described as:

$$EC_{XY} = \frac{[X] \cdot [Y]}{[XY]} \qquad (9.52)$$

This equilibrium constant can be established by measuring the concentrations (C_{X1} and C_{X2}) in the headspace of two solutions with different concentrations of X and Y but having the same total volume of the solutions. Assuming that the distribution constant (partition coefficient) of the volatile reactant remains constant in the studied concentration range, then one can express the equilibrium constant of the reaction in the following way [29]:

$$EC_{XY} = \frac{C_{X2} \cdot [Y]_2 - C_{X1} \cdot [Y]_1}{C_{X1} - C_{X2}} \cdot \frac{K_X \cdot V_S}{K_X \cdot V_S + V_G} \qquad (9.53)$$

where V_S and V_G are the respective volumes of the liquid and gas phase in the vial, C_X is the concentration of the volatile reactant in the headspace, K_X is the partition coefficient of the volatile reactant in the vial, and $[Y]$ is the concentration of the nonvolatile reactant in the vial.

This method was also applied to determine the stability constants (SC_{XM}) of coordination complexes of substance M with the volatile ligand X:

$$SC_{XM} = \frac{[M] \cdot [X]}{[MX]} = \frac{A_1 - A_2}{A_2} \cdot \frac{K_X \cdot V_S + V_G}{K_X \cdot V_S \cdot [M]_o} \qquad (9.54)$$

where A is the peak area of the volatile ligand and $[M]_o$ is the initial concentration of the coordination substance. In addition, this method was applied to express the basicity of volatile compound B by the determination of the ionization constant IC_{BH^+} of the basic conjugate BH^+ from the area ratio AB_1/AB_2 of compound B in two solutions with different concentrations and pH [30,31]:

$$IC_{BH^+} = \frac{[B] \cdot [H^+]}{[BH^+]} = \frac{(A_{B1}/A_{B2}) \cdot [H^+]_1 - [H^+]_2}{1 - A_{B1}/A_{B2}} \qquad (9.55)$$

9.6 Gas–Solid Systems

Essentially the technique described earlier for gas–liquid systems can be applied for gas–solid systems to investigate adsorption effects and diffusion processes.

9.6.1 Determination of Adsorption Isotherms

Similar to the method for the determination of the partition coefficients, it is also possible to measure adsorption isotherms [11,16] using, for example, the VPC method. Increasingly larger aliquots of the volatile compounds are introduced into a series of two parallel sets of vials, one containing a constant amount of the adsorbent, while the other is empty, thus having always a pair with the same amount of the volatile compound: the resulting difference in the area values corresponds to the adsorbed amount.

In the calculation it is necessary in principle to make corrections for the sample volume because of the difference in the headspace volume of the pair of vials compared (vial with adsorbent vs. empty vial); also, the porosity of the adsorbent should be considered because this volume also contributes to the free gas volume of the vial. In practice, however, the amount of the adsorbent under investigation seldom exceeds 100 mg (0.1 mL) in a 22.3 mL vial, and the difference of $<1\%$ can be ignored.

Adsorption isotherms can be established by plotting the corresponding peak area ratios A_i/A_i^o instead of the usual relative pressures p_i/p_i^o against the adsorbed amount in millimoles per gram. An example is shown in Figure 9-7 for the adsorption of ethanol on activated charcoal and silica gel, at 60 °C [11]. From the adsorption isotherms the corresponding thermodynamic functions such as adsorption coefficients, adsorption enthalpies, and specific surface area can be obtained [32]. HS-GC is also helpful in the investigation of other selective sorption properties or the investigation of catalysts (see, e.g., refs. 33–35).

HS-GC is also a useful technique in environmental applications — for example, in studying the adsorption of organic compounds on soil or similar materials. Thus, Figure 9-8 shows adsorption isotherms of trichloroethylene on dry and wet peat and of toluene on dry and wet clay. These materials were used as standards for soil investigations: as shown here, the dry peat had a very high adsorptivity and wet peat still had a reasonable one. On the other hand, wet clay corresponding to natural humidity had practically no adsorptivity for toluene, and only the dry material was slightly more adsorptive.

9.6.2 Determination of the Rate of Release of a Volatile Analyte

Adsorption equilibrium is established by diffusion processes from both sides of the two-phase system. It is thus feasible to study such diffusion effects by using kinetic measurements in the same way as applied for the gas phase addition technique (cf. Section 5.4.2 and Figure 5-10). Of some interest here is the determination of the *rate of release* (*RR*) of a volatile analyte from a solid sample, particularly if it is a toxic compound.

At the start of the equilibration process there is no analyte in the gas phase of the vial: its concentration will slowly build up there until equilibrium is reached. This buildup can be presented by analyzing the headspace of a number of vials thermostatted to increasing times and plotting the obtained peak heights against the thermostatting time. For the shape of such a plot, see Figure 4-1.

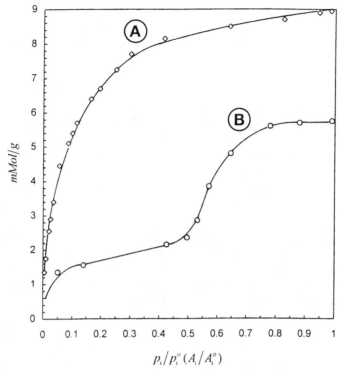

Figure 9-7. Adsorption isotherms of ethanol on activated charcoal, (*A*) and silica gel, (*B*) at 60 °C [11]. The relative pressures p_i/p_i^o are expressed as the peak area ratios A_i/A_i^o. Ordinate: adsorbed amount of ethanol per gram of adsorbent. Activated charcoal: "for chromatography," 20/30 mesh, E. Merck; silica gel: narrow pore, Davison grade 12, 60/80 mesh.

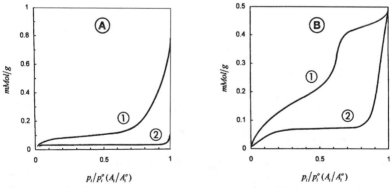

Figure 9-8. Adsorption isotherms at 80 °C (A) Toluene on dry and wet clay. (B) Trichloroethylene on dry and wet peat. The relative pressures p_i/p_i^o are expressed as the peak area ratios A_i/A_i^o. Ordinate: adsorbed amount of toluene and trichloroethylene per gram of adsorbent. *1* = dry adsorbent, dried at 150 °C, *2* = wet adsorbent (the humidity of the wet peat was about 50%).

At the beginning of the equilibration process, the analyte diffuses only in one direction: from the sample into the gas phase. In this period, the relationship between peak height H_t and thermostatting time t increases linearly:

$$H_t = a \cdot t \tag{9.56}$$

The plot will flatten when the gas phase concentration builds up and diffusion from the gas phase back into the sample begins. Finally, at equilibrium both diffusion processes are equal, resulting in a constant peak height, now independent of the thermostatting time.

The early, linear portion of the equilibration plot can be used to estimate the rate of release of the analyte from the solid sample. What we need is the relationship between amount of analyte W_i and peak height H_t:

$$W_i = f \cdot H_t \tag{9.57}$$

in which the factor f is similar to the response factor in general gas chromatography, and can be established by, for example, the analysis of an external vapor standard of the analyte. Thus, we can substitute the corresponding values of the amount of the analyte for the measured peak height. The linear portion of the plot will correspond to the

$$W_i = a' \cdot t \tag{9.58a}$$

relationship, the slope of which (determined by linear regression) has the dimension of amount/time:

$$a' = W_i/t \tag{9.58b}$$

We relate this to the total amount of sample present W_{sample} and express as the *rate of release* RR of the analyte from the sample:

$$RR = \frac{W_i/t}{W_{sample}} \tag{9.59}$$

The rate of release can also be related to the surface of the sample S_{sample}:

$$RR = \frac{W_i/t}{S_{sample}} \tag{9.60}$$

The dimension of the term described by eq. 9.59 is

$$\left[\frac{\text{(mass of analyte)/time}}{\text{mass of sample}} \right]$$

while the dimension of the term described in eq. 9.60 is

$$\left[\frac{\text{(mass of analyte/time}}{\text{surface area of sample}} \right]$$

We now give an example for such a measurement. Poly(ethylene terephthalate) (PET) is an important plastic in the production of bottles and packaging films. A thermal degradation product that may be formed during the polymerization reaction and melt-processing of PET is acetaldehyde (AA), and thus its determination is of general importance. According to Dong et al. [36], this determination can be carried out by HS-GC with freeze-ground PET pellets. On the other hand, since PET is used as food packaging material, it might also be of interest to know how much acetaldehyde is released into the surrounding atmosphere. This can be determined by measuring the rate of release from PET granules in a headspace vial by HS-GC.

Figure 9-9 plots the amount of AA released (determined by an external vapor standard of AA) as a function of the thermostatting time, at three different temperatures, and Table 9-4 gives the pertinent data for 135 °C. The external vapor standard was prepared by injecting 1 μL of acetaldehyde into an empty vial and analyzing it under the same conditions. The sample always consisted of 1.0 g PET granules. As seen, at 120 °C the plot is linear up to 120

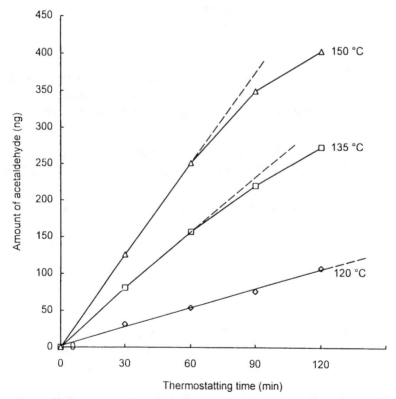

Figure 9-9. Release of acetaldehyde from poly(ethylene terephthalate) granules as a function of the thermostatting time, at three different temperatures. Values of the correlation coefficient r for the linear portions of the plots and the calculated rate of release RR: 120 °C: $r = 0.9876$, $RR = 0.84$ (ng/min)/g; 135 °C: $r = 0.9986$, $RR = 2.63$ (ng/min)/g; 150 °C: $r = 1.0000$, $RR = 4.19$ (ng/min)/g.

Table 9-4 Release of acetaldehyde (AA) from poly(ethylene terephthalate) pellets (1.0 g) as a function of the thermostatting time, at 135 °C

Thermostatting time t (min)	Peak height H_t (mm)	Corresponding amount of AA W_i (ng)	Rate of release RR (ng/min)
0^+	0.0	0.0	
30^+	9.0	80.7	2.69
60^+	17.5	157.0	2.62
90	24.5	219.8	
120	30.5	273.6	
			mean: 2.66
Linear regression[†]			
Correlation coefficient r			0.99986
Slope a'			2.62 ng/min

* Response factor: 8.97 ng AA per millimeter of peak height.
[†] Of the first three values (marked with +), according to eq. 9.58a.

minutes thermostatting time, and up to 60 minutes at 135 °C and 150 °C; thus, we can carry out linear regression analysis of these data according to eq. 9.58a. In Table 9-4 we also give RR data calculated for each individual measurement: their mean agrees well with the value obtained from linear regression analysis.

9.7 Validation of the Headspace Instrumentation: Investigation of Detector Linearity and Detection Limit

We have discussed various techniques and possibilities for quantitative headspace analysis. Once the method has been established, it must be validated. An essential part of method validation is the determination of the instrument performance. This includes the linearity of the working range and the detection limit.

9.7.1 Definitions

The *linear range of a detector* is defined as the range in which the detector's sensitivity [(peak area (A_i) per analyte amount (W_i)] is constant [37]:

$$A_i/W_i = \text{const.} \tag{9.61}$$

In general, the linear range is established by plotting detector sensitivity against the injected analyte amount and defining the range in which the values differ by less than a specified variation ("window"), usually $\pm 5\%$. The preparation of the many standards, required for controlling a wide concentration range, includes the error of manual sample preparation. Therefore the result of such a measurement series does not necessarily represent the true instrument precision.

The linear range of a detector can also be established with help of headspace–gas chromatography, using a set of MHE measurements from a vapor standard. After all, in MHE we obtain the peak area for successively smaller amounts of the analyte present. Since we know the absolute amount of the analyte corresponding to the highest peak from preparing the vapor standard, we can prepare any desired lower concentration by the stepwise automatic dilution procedure. Thus the detector linearity can automatically be controlled, if not over the whole range, at least for the working range of interest. No problems with unreproducible sample preparation are involved, since a single sample, the vapor standard with the highest concentration, is automatically diluted by the stepwise MHE procedure.

The minimum detectable limit DL of a detector is considered to be the amount of analyte that would have a peak height equal to x times the noise level N of the detector. It is calculated from the values of a small peak approaching the detection limit, assuming proportionality between peak height and analyte amount:

$$DL = \frac{x \cdot N}{H_i} \cdot W_i \tag{9.62}$$

where H_i is the height of the peak corresponding to the amount of W_i. The detectability can also be expressed with respect to the carbon content of the analyte molecule. IUPAC [37] and instrument specifications usually consider $x = 2$, but other values (e.g., $x = 5$) have been used in chromatographic practice.

In eq. 9.62 DL is expressed in mass, usually in micrograms, nanograms, or picograms. In mass–flow sensitive detectors (e.g., the flame-ionization detector) the mass flow rate of the analyte entering the detector (in pg/s) is used instead of its mass:

$$DL/t = \frac{x \cdot N}{H_i \cdot w_h} \cdot W_i \tag{9.63}$$

where w_h is the peak width at half-height, expressed in units of time.

In MHE if we carry out a nine-point measurement, we can adjust the starting concentration of the vapor standard so that the last peak will be close to the detection limit. Therefore, its values can be used to calculate DL.

In Section 9.7.2 we illustrate the possibility of such detector tests by the example of a flame-ionization detector. Toluene was used as the analyte; the sample was prepared by adding 2.5 μL of a 1.0 vol% solution of toluene in acetone to a headspace vial of 22.3 mL volume and using the total vaporization technique, by thermostatting the vial at 100 °C for 30 minutes. The density of toluene is 0.866 g/mL; thus the added solution contained 21.65 μg toluene and its concentration in the gas phase present in the vial was 0.971 μg/mL.

Table 9-5   MHE measurement of the toluene vapor standard used to test the linearity of the working range of a flame-ionization detector*

	Peak area (counts)		$\dfrac{|A_{i(theor)} - A_{i(meas)}|}{A_{i(theor)}}$
i	Measured	Theoretical[†]	
1	2,369,497	2,347954	0.0092
2	1,086,966	1,043,021	0.0421
3	464,134	463337	0.0017
4	201,158	205,826	0.0227
5	87,970	91,433	0.0379
6	38,761	40,617	0.0457
7	17,982	18,043	0.0034
8	8,311	8,015	0.0369
9	3,640	3,561	0.0222
Sum of the absolute relative diffrerences:			0.2218
Linear regression			
Correlation coefficient r	−0.99990		−1.00000
Slope q	−0.8242		
$Q = e^{-q}$	0.4442		
Intercept A_1^*	2,347,954		
ΣA_i^*	4,224,655		

*   Conditions: 50 m × 0.32 mm I.D. open-tubular column coated with bonded phenyl (5%) methyl silicone phase; film thickness: 1 μm. Column temperature: 70 °C. Sample and thermostatting is specified in the text. Vial pressure $\Delta p = 120$ kPa.
†   Calculated from the linear regression equation, assuming that $r = -1.00000$.

We carried out a nine-step MHE measurement, the results of which are listed in Table 9-5 under "Peak area measured." The correlation coefficient indicates excellent linearity in the studied range.

9.7.2   Linear Range of the Detector

First, the linear range corresponding to this measurement series is established. The amount of toluene corresponding to the first peak area can be estimated in the following way: the actual gas volume transferred from the headspace vial into the column is calculated as described in Section 3.5.6, correcting the flow rate values to dry gas conditions and to the temperature of the vial. The calculation is straightforward, corresponding to eqs. 3.9 and 3.7, respectively, and to the general chromatographic relationships concerning the correction of flow rates [38]:

$$F_{c,o} = F_a \cdot \frac{T_V}{T_a} \cdot \frac{p_a - p_w}{p_a} \tag{3.9}$$

$$F_i = \frac{p_a}{p_i} \cdot F_{c,o} w \tag{3.7a}$$

where F_a is the flow rate measured with a soap-bubble flowmeter at column outlet and atmospheric conditions (pressure p_a and temperature T_a); in addition, $F_{c,o}$ is the value of F_a corrected to dry gas conditions (p_w = vapor pressure of water at T_a) and vial temperature T_V; and F_i is the flow rate at column inlet pressure ($p_i = \Delta p + p_a$). The following values were measured:

$$F_a = 2.8 \text{ mL/min}$$

$$p_a = 96.42 \text{ kPa}$$

$$\Delta p = 120.0 \text{ kPa}$$

$$p_i = 96.42 + 120.0 = 216.42 \text{ kPa}$$

$$T_a = 20\,°C = 293.16 \text{ K}$$

$$T_V = 100\,°C = 373.16 \text{ K}$$

$$p_w \text{ at } 20\,°C = 2.332 \text{ kPa}$$

$$F_{c,o} = 2.8 \cdot \frac{373.16}{293.16} \cdot \frac{96.42 - 2.332}{96.42} = 3.478 \text{ mL/min}$$

$$F_i = 3.478 \cdot \frac{96.42}{216.42} = 1.55 \text{ mL/min}$$

The sample transfer time was $t = 0.06$ minutes. Thus, the volume of headspace gas transferred in the first injection was

$$V_{gas} = F_i \cdot t = 1.55 \times 0.06 = 0.093 \text{ mL} = 93.0 \,\mu L$$

As given earlier, the toluene concentration in the headspace vial was $0.971 \,\mu g/mL$: thus, this transferred volume corresponds to

$$\frac{93.0 \times 0.971}{1000} = 0.0903 \,\mu g = 90.3 \text{ ng toluene}$$

Based on the proportionality of peak area versus amount, the amount corresponding to the ninth analysis can be calculated from the respective peak area, using the theoretical values of A_1 and A_9 (see Table 9-5) and the amount of toluene transferred in the first measurement (90.3 ng):

$$\frac{3561}{2{,}347{,}954} \cdot 90.3 = 0.137 \text{ ng toluene}$$

Thus, the investigated analyte range was 90.3 ng/0.137 ng = 659 : 1.

Two remarks are necessary here. First, it is important to realize that such a calculation can in general be carried out only in the case of splitless injection into an open-tubular column (see Figure 3-12II), while in the split mode (see Figure 3-12I) the headspace sample is diluted in the injector by the carrier gas. The whole evaluation depends on the possibility of accurately establishing the amount of the analyte present in the sample transfer in the first step of the

MHE procedure. For this reason it is essential to determine the exact volume of gas transferred into the column. With modern programmed, pressure-controlled instruments, however, where the carrier gas flow to the injector can be stopped or at least the column inlet pressure p_i strongly reduced during sample transfer, it is still possible to work in the split mode because dilution of the headspace sample is avoided. It is not even necessary to determine the split flow for this calculation.

Second, this example shows also that one nine-step MHE measurement cannot provide evaluation of the full linearity range of an FID, which is generally considered to be $10^7 : 1$. The range we investigated was only $659 : 1$, and if we extend it to the minimum detectable limit (see Section 9.7.4), it covers the range of $90.3 \, \text{ng}/0.014 \, \text{ng} = 6.5 \times 10^3 : 1$. To evaluate the whole linear range, one would need two or more overlapping MHE measurements, starting at different gas phase concentrations. However, the wide linear range of an FID cannot be utilized with open-tubular columns, with their limited sample capacity. Therefore the single range investigated here is already sufficient for method validation and for the investigation of the detector, particularly in connection with the evaluation of the minimum detectability, as outlined shortly.

If one is investigating a detector with a shorter linear range, it may be that the concentration of the standard is too high in the gas phase of the vial at the beginning of the MHE measurement series. This would immediately be evident from the linear regression data, which would show a poor correlation coefficient, and if the usual MHE plot is prepared, it will demonstrate the nonlinearity in its earlier part. A good example appears in Figure 9-10, which shows the MHE plot from investigating the linearity of an electron–capture detector, using chloroform as the analyte. The starting chloroform concentration in the vial at MHE step 1 was $1.54 \, \text{ng/mL}$ and this was clearly too high. In the ninth MHE step the concentration in the vial was $9.65 \, \text{pg/mL}$, which is close to the minimum detectability of this type of detector.

In such a case one has two possibilities: preparing a new standard with a smaller amount of the analyte, or carrying out calculation only for the linear part of the MHE plot. That the second option is possible is demonstrated in the plot shown in Figure 9-10: the correlation coefficient for the fifth to ninth steps is $r = 0.99997$, which is very good.

9.7.3 Precision of the Range

Using the linear regression analysis of the MHE, we can also define a term for the precision $P\%$ of the range. This can be established by first calculating the "theoretical peak area" $A_{i(\text{theor})}$ and then the absolute relative differences between theoretical and measured $A_{i(\text{meas})}$ peak area for each step:

$$\frac{|A_{i(\text{theor})} - A_{i(\text{meas})}|}{A_{i(\text{theor})}}$$

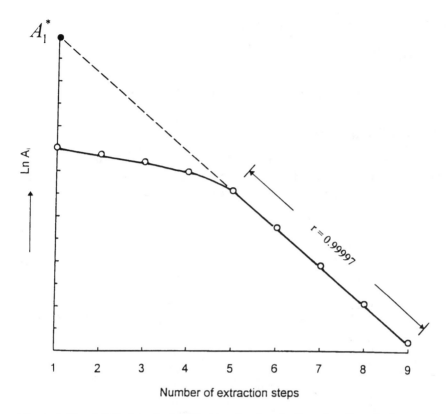

Figure 9-10. MHE plot obtained in investigating the linearity of an electron capture detector. *HS conditions*: Sample: $2\,\mu L$ of a chloroform solution $(600\,\mu g/mL)$, thermostatted at $80\,°C$. *GC conditions*: Column: $50\,m \times 0.32\,mm$ I.D. fused-silica, open-tubular column, coated with bonded phenyl (5%) methyl silicone stationary phase; film thickness: $2\,\mu m$. Sample transfer: 1 minute by cryofocusing, splitless. Column temperature: $40\,°C$ isothermal for 5 minutes, then programmed at $5\,°C/min$ to $120\,°C$ and then at $20\,°C/min$ to $150\,°C$. Carrier gas: Helium, $130\,kPa$; makeup gas to the ECD: argon-methane, $50\,mL/min$. ECD temperature: $350\,°C$. The starting chloroform concentration in the vial at MHE step 1 was $1.54\,ng/mL$, corresponding to the extrapolated intercept value A_1^*, while the final concentration at step 9 was $9.65\,pg/mL$. The correlation coefficient in the linear part of the detector response is $r = 0.99997$.

Averaging these data the precision of the range (in percent) can be calculated:

$$P\% = \frac{\sum\limits_{i=1}^{n} \dfrac{|A_{i(\text{theor})} - A_{i(\text{meas})}|}{A_{i(\text{theor})}}}{n} \tag{9.64}$$

Table 9-5 gave values of the absolute relative differences and their sum for the investigation of the linearity of a flame-ionization detector, using toluene as the test substance. The result of the calculation is:

$$P\% = \frac{0.2218 \times 100}{9} = \pm 2.5\%$$

which is well within the generally specified $\pm 5\%$ window.

9.7.4 Minimum Detectability

Figure 9-11 shows the actual peak obtained in the ninth measurement in testing the linearity of an FID (Table 3-5). As given earlier, this peak with the peak height H_9 corresponded to $W_9 = 137\,\text{pg}$ toluene. The characteristics of

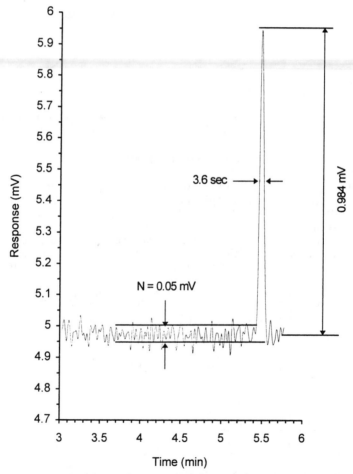

Figure 9-11. Toluene peak from the ninth MHE measurement, used to calculate the minimum detectable limit of a flame-ionization detector. $N = $ noise level.

this peak were:

peak height $(H_9) = 0.984\,mV$

peak width at half-height $(w_h) = 3.6$ seconds

noise level $(N) = 0.05\,mV$

If we consider the detection limit (DL) as a peak with a height twice the noise level, the corresponding amount can be calculated based on the proportionality of peak height and amount:

$$DL = \frac{2N}{H_9} \cdot W_9 = \frac{2 \times 0.05}{0.984} \cdot 137 = 13.92\,pg$$

and since the FID is a mass-flow-sensitive detector, we should express the detection limit as amount corresponding to unit time:

$$DL/t = \frac{13.92}{3.6} = 3.9\,pg/s \text{ toluene.}$$

Since 91.24% of the molecular weight of toluene represents carbon, this value corresponds to $3.6\,pgC/s$. In general, the detection limit of flame-ionization detectors is specified as $3-5\,pgC/s$. Thus, the measurement showed that the minimum detectability of the tested detector is within the required range.

References

1. D. C. Legget, *J. Chromatogr.* **133**, 83–90 (1977).
2. K. Schoene and J. Steinhanses, *Fresenius Z. Anal. Chem.* **309**, 198–200 (1981).
3. K. Schoene, K. W. Böhmer, and J. Steinhanses, *Fresenius Z. Anal. Chem.* **319**, 903–906 (1984).
4. A. Hussam and P. W. Carr, *Anal. Chem.* **57**, 793–801 (1985).
5. J. E. Woodrow and J. N. Seiber, *J. Chromatogr.* **455**, 53–65 (1988).
6. B. Kolb and C. Welter, presentation of the sixth COLACRO, Latin American Congress on Chromatography, January 21–25, 1996, Caracas, Venezuela.
7. R. C. Weast (editor), *Handbook of Chemistry and Physics*, 57th ed., 1976–77, CRC Press, Cleveland, OH; D-204.
8. B. Kolb, *CZ-Chem. Tech.* **1**, 87–91 (1972).
9. B. Kolb, *J. Chromatogr.* **112**, 287–295 (1975).
10. J. H. Hildebrand and R. L. Scott, *The Solubility of Nonelectrolytes.* Dover, New York, 1964; p. 181.
11. B. Kolb, P. Pospisil, and M. Auer, *Ber. Bunsenges. Phys. Chem.* **81**, 1067–1070 (1977).
12. H. Hachenberg and A. P. Schmidt, *Gas Chromatographic Headspace Analysis*, Heyden & Son, London, 1977; Part 2, pp. 81–116.
13. U. Weidlich , J. Berg, and J. Gmehling, *J. Chem. Eng. Data*, **31**, 313–317 (1986).
14. I. M. Balashova, L. V. Mokrushina and A. G. Morachevsky, presentation at the International Conference Chisa-90, Prague, Czechoslovakia, 1990, p. 67.
15. I. M. Balashova, L. V. Mokrushina and A. G. Morachevsky, *Theor. Found. Chem. Technol* (in Russian), *TOXT*, V. 3, N4, 366–382 (1996).
16. B. Kolb, in B. Kolb (editor), *Applied Headspace Gas Chromatography*, Heyden & Son, London, 1980; pp. 1–11.
17. B. V. Ioffe and A. G. Vitenberg, *Headspace Analysis and Related Methods in Gas Chromatography*, Wiley, New York, 1984; pp. 234–263.

18. K. Schoene, J. Steinhanses, and A. König, *J. Chromatogr.* **455**, 67–75 (1988).
19. K. Schoene and J. Steinhanses, *Fresenius Z. Anal. Chem.* **321**, 538–543 (1985).
20. G. A. Robbins, S. Wang, and J. D. Stuart, *Anal. Chem.* **65**, 3113–3118 (1993).
21. L. Rohrschneider, *Anal. Chem.* **45**, 1241–1247 (1973).
22. B. Kolb, C. Welter, and C. Bichler, *Chromatographia*, **34**, 235–240 (1992).
23. J. J. Jalbert and R. Gilbert, Conference Record of the 1994 IEEE International Symposium on Electrical Insulation, Pittsburgh, PA, June 5–8, 1994; pp. 123–129.
24. L. S. Ettre, C. Welter, and B. Kolb, *Chromatographia*, **35**, 73–84 (1993).
25. C. McAuliffe, U.S. Patent 3,759,086 (1973).
26. A. N. Marinichev and B. V. Ioffe, *J. Chromatogr.* **454**, 327–334 (1988).
27. K. Schoene and J. Steinhanses, *Monatsh. Chem. (Wien)* **117**, 1927–1939 (1986).
28. F. F. Vincieri, G. Mazzi, N. Mulinacci, P. Pappini, and N. Gelsomini, *Pharm. Acta Helv.* **63**, 282–286 (1988).
29. In ref. 17, pp. 242–244.
30. In ref. 17, pp. 247–254.
31. A. G. Vitenberg, Z. St. Dimitrova, and B. V. Ioffe, *J. Chromatogr.* **171**, 49–54 (1979).
32. J. B. Pausch, *J. Chromatogr. Sci.* **22**, 161–164 (1984).
33. H. Hachenberg, H. Baltes, E. G. Schlosser, H. Littner, E. J. Leupold, and E. Frost, *Erdö Kohle-Erdgas-Petrochem., Brennstoffchem.* **36**, 418–422 (1983).
34. K. Schoene, J. Steinhanses, and W. Wienand, *J. Colloid Interface Sci.* **91**, 595–597 (1983).
35. K. Schoene, J. Steinhanses, and A. König, *J. Chromatogr.* **514**, 279–286 (1990).
36. M. Dong, A. H. DiEdwardo, and F. Zitomer, *J. Chromatogr. Sci.* **18**, 242 (1980).
37. L. S. Ettre (compiler), Nomenclature for Chromatography Issued by the Analytical Chemistry Division of the IUPAC, *Pure Appl. Chem.* **65**(4), 819–872 (1993).
38. L. S. Ettre and J. Hinshaw, *Basic Relationships of Gas Chromatography*, Advanstar, Cleveland, OH, 1993; pp. 35–36.

Compound Index:
Analytes, Samples, Solvents, and Reagents

Entries are arranged alphabetically by *parent compounds*; derivatives (i.e., bromo-, ethyl-, etc.) are treated as subentries. Compounds used as solvents and displacers or modifiers are marked with suffix s. Compounds, used as reagents or analyzed as a derivative resulting from a chemical reaction are marked with suffix r.

Subject Index